Jörg Pfeifle

Terabit-Rate Transmission Using Optical Frequency Comb Sources

Karlsruhe Series in Photonics & Communications, Vol. 20
Edited by Profs. C. Koos, W. Freude and S. Randel

Karlsruhe Institute of Technology (KIT)
Institute of Photonics and Quantum Electronics (IPQ)
Germany

Terabit-Rate Transmission Using Optical Frequency Comb Sources

by
Jörg Pfeifle

KIT Scientific Publishing

Dissertation, Karlsruher Institut für Technologie (KIT)
KIT-Fakultät für Elektrotechnik und Informationstechnik, 2016

Impressum

KIT Scientific Publishing

Karlsruher Institut für Technologie (KIT)
KIT Scientific Publishing
Straße am Forum 2
D-76131 Karlsruhe

KIT Scientific Publishing is a registered trademark of Karlsruhe Institute of Technology. Reprint using the book cover is not allowed.

www.ksp.kit.edu

Print on Demand 2017 – Gedruckt auf FSC-zertifiziertem Papier

ISSN 1865-1100
ISBN 978-3-7315-0640-9
DOI 10.5445/KSP/1000066936

Terabit-Rate Transmission Using Optical Frequency Comb Sources

Zur Erlangung des akademischen Grades eines

DOKTOR-INGENIEURS

Von der Fakultät für Elektrotechnik und Informationstechnik
des Karlsruher Instituts für Technologie (KIT)

genehmigte

DISSERTATION

von

Dipl.-Ing. Jörg Pfeifle

geboren in

Tübingen, Deutschland

Tag der mündlichen Prüfung: 29.07.2016
Hauptreferent: Prof. Dr. Christian Koos
Korreferenten: Prof. Dr.-Ing. Dr. h.c. Wolfgang Freude
Prof. Dr. Jeremy Witzens

“Hell, there ain’t no rules around here!
We are tryin’ to accomplish somep’n!”
– Thomas A. Edison

Table of Contents

Abstract (German)

Die durchschnittliche Wachstumsrate des Datenverkehrs über optische Glasfasernetzwerke beträgt 23% pro Jahr [1]–[3]. Es wird erwartet, dass das Jahresgesamtvolumen im Jahr 2016 ein Zettabyte übersteigen wird. Hierbei wird zwischen Metro- und Langstrecken-Netzwerken unterschieden, wobei seit 2014 der Metro-Anteil größer ist und doppelt so schnell wächst wie der Langstrecken-Anteil. Hauptgrund ist die zunehmende Bedeutung der Distribution großer Dateien, z.B. Videos.

Die optischen Kommunikationssysteme, die diese Datensignale transportieren, müssen verbessert werden, um auch in Zukunft dem wachsenden Datenvolumen gerecht werden zu können. Es gibt drei Möglichkeiten, um die Datenrate der optischen Kommunikationssysteme zu erhöhen. Erstens kann die Anzahl der Bits pro übertragenem Symbol, unter Verwendung von höherwertigen Modulationsformaten, erhöht werden. Dies resultiert jedoch in höheren Anforderungen an die Einzelkomponenten des Übertragungssystems, sowie an das Signal-zu-Geräusch-Leistungsverhältnis, was letztlich die Reichweite des Kommunikationssystems limitiert. Zweitens kann die Symbolrate erhöht werden, was allerdings durch die Bandbreite der verfügbaren Elektronik begrenzt ist. Die dritte und verbleibende Möglichkeit ist, die Anzahl der parallelen Datenkanäle zu erhöhen. Üblicherweise werden hierzu viele unterschiedliche Trägerfrequenzen verwendet, was als dichtes Wellenlängenmultiplex-Verfahren (engl.: Dense Wavelength-Division Multiplexing, DWDM) bekannt ist.

In heutigen DWDM-Systemen werden pro Leitungskarte (engl.: Line Card) Daten mit einer Rate von 100 Gbit/s übertragen. In jeder Leitungskarte werden dabei die Daten auf einen einzelnen Träger moduliert. Getrieben durch das weiterhin steigende Datenvolumen, soll die Datenrate pro Leitungskarte weiter erhöht werden. An den technischen Möglichkeiten, um eine Datenrate von 400 Gbit/s und 1 Tbit/s zu erreichen, wird intensiv geforscht. Nach bisherigen Erkenntnissen ist der vielversprechendste Ansatz, mehrere Träger für ein Datensignal zu verwenden, diese zu einem sogenannten Superkanal zu gruppieren und in eine Leitungskarte zu integrieren. Innerhalb dieser Superkanäle können die spektralen Schutzintervalle zwischen den zugehörigen Wellenlängenkanälen minimiert werden. Für die Hersteller der Leitungskarten bedeutet diese Entwicklung, dass die Funktionalität von mehreren Leitungskarten in eine einzelne Leitungskarte integriert werden muss. Hierzu müssen die einzelnen Komponenten kleiner, preiswerter und energiesparender werden. In der vorliegenden Arbeit betrachten wir die Verwendung eines optischen Frequenzkammes als Vielträgerquelle, um eine Vielzahl einzelner Laser zu ersetzen.

Der Vorteil eines optischen Frequenzkammes ist, dass das komplette Frequenzraster durch zwei Parameter, Mittenfrequenz und freier Spektralbereich, vollständig bestimmt ist. Dadurch entfällt die individuelle Stabilisierung der einzelnen Laserdioden, und es verringert sich somit potentiell Platz- und Energiebedarf. Zusätzlich ermöglichen die perfekt äquidistanten Kammlinien die Minimierung spektraler Schutzintervalle, da die relative Verschiebung der Trägerfrequenzen zueinander minimal ist.

In dieser Arbeit werden drei Frequenzkammquellen untersucht, welche als Mehrträgerquelle in eine Leitungskarte integriert werden könnten. Wir zeigen, dass jede der untersuchten Frequenzkammquellen für einen Datenstrom von mehr als 1 Tbit/s geeignet ist. Die erste Frequenzkammquelle ist ein Laser, in dem durch starke Modulation des Injektionsstroms ein Frequenzkamm erzeugt wird (engl.: Gain-Switched Comb Source, GSCS). Diese Kammquelle besticht durch außerordentliche Stabilität und Flexibilität. Es wird gezeigt, dass trotz der geringen Anzahl von etwa 10 Kammlinien die Übertragung von Datenraten im Tbit/s-Bereich über eine Entfernung von mehreren hundert Kilometern möglich ist. Die zweite Kammquelle ist eine modengekoppelte Quantendraht-Laserdiode (engl.: Quantum-Dash Mode-Locked Laser Diode, QD-MLLD). Diese Laser sind besonders kompakte Quellen und emittieren mehr Kammlinien als die GSCS, allerdings mit großem Phasenrauschen. Es werden Möglichkeiten aufgezeigt, um trotz dieser Einschränkung eine Tbit/s-Datenverbindung über 75 km aufzubauen. Die Dritte ist eine besonders breitbandige Kammquelle, bei der die Kammlinien durch nichtlineare Prozesse in einem Mikroresonator erzeugt werden. Diese Frequenzkämme überspannen mehrere Telekommunikationsbänder und sind besonders für Multi-Tbit/s-Datenübertragung geeignet.

Kapitel 1 gibt eine allgemeine Einführung in die Datenübertragung mit Wellenlängenmultiplex-Verfahren sowie einen Überblick zum Stand der Technik. Die Verwendung von Frequenzkämmen als Träger für Superkanäle wird motiviert, und die dazugehörige Transmitter-Architektur wird vorgestellt. Anschließend werden die in dieser Arbeit untersuchten Frequenzkammquellen sowie der Aufbau eines künftigen Tbit/s-Sende-Moduls beschrieben.

Kapitel 2 fasst die physikalischen und technischen Grundlagen von Freuqenzkammquellen zusammen und diskutiert die Anforderungen an die einzelnen Kammlinien für die Verwendung in der kohärenten Datenübertragung.

Kapitel 3 bespricht die kohärente Datenübertragung mit einer GSCS. Es enthält eine Einführung in diese Technik der Kammerzeugung sowie den experimentellen Aufbau, mit dem ein Superkanal-Sender und -Empfänger emuliert wird. Durch den variablen Linienabstand dieser Kammquelle konnten verschiedene Experimente mit mehreren Symbolraten und Modulationsformaten durchgeführt werden. Im besten Fall werden Daten mit einer Rate von 1.9 Tbit/s erzeugt und über 300 km übertragen.

Kapitel 4 erläutert die Datenexperimente, die mit einer QD-MLLD durchgeführt wurden. Diese Kammquelle sticht besonders durch ihre Kompaktheit hervor. Allerdings weisen die Kammlinien ein besonders hohes Phasenrauschen auf. Zwei Möglichkeiten werden vorgestellt, um trotz des hohen Phasenrauschens ein Tbit/s-Datensignal mit kohärenter Sende- und Empfangstechnologie zu übertragen. Zum einen eine Technik der Phasenrauschkompensation, welche das Phasenrauschen der Kammlinien reduziert, bevor die Daten auf die Träger moduliert werden. Hierbei wird eine Datenrate von 1 Tbit/s über 75 km übertragen. Zum anderen wird eine Datenübertragung von 1.6 Tbit/s über 75 km unter Verwendung von Selbst-homodyn-Empfang demonstriert. In diesem Fall werden die zu den Kammlinien gehörenden Felder aufgeteilt: Ein Teil dient als Träger des Datenstroms, während der andere Teil zusammen mit dem Datenstrom auf orthogonalen Polarisationen zum Empfänger übermittelt wird. Auf der Empfängerseite werden die nicht-modulierten Kammlinien dann als Lokaloszillatoren mit zu den Datenträgern korreliertem Phasenrauschen verwendet.

Kapitel 5 diskutiert Datenübertragungsexperimente, für die ein in einem Mikroresonator erzeugter Frequenzkamm verwendet wird. Die eingesetzten Materialien und Resonator-Geometrien sowie die Dynamik der Kammerzeugung werden erläutert. Es stellt sich heraus, dass die Eigenschaften der erzeugten Frequenzkämme wesentlich durch das Verfahren zur Einstellung der Parameter des Pumplasers sowie das Dispersionsprofil des Mikroresonators beeinflussbar sind. Es werden mehrere Datenübertragungsexperimente vorgestellt, bei denen unterschiedliche Zustände dieser sogenannten Kerr-Kämme verwendet wurden. Die höchste

Datenrate von 19.7 Tbit/s konnte mit einem Solitonen-Kammzustand erreicht und über 75 km Glasfaser übertragen werden.

Kapitel 6 fasst die wichtigsten Ergebnisse der vorliegenden Arbeit zusammen und zieht einen Vergleich zu anderen kohärenten Datenexperimenten, die eine Frequenzkammquelle verwenden. Darüber hinaus gibt das Kapitel einen Ausblick über Möglichkeiten, die untersuchten Kammquellen weiter zu verkleinern und zu verbessern.

Die im Rahmen dieser Dissertation entstandenen Ergebnisse wurden in mehreren wissenschaftlichen Zeitschriften publiziert und auf fünf internationalen Konferenzen präsentiert. Eine Auflistung aller Publikationen ist auf Seite 167 ff. zu finden.

Preface

Global internet traffic is growing at a compound annual growth rate of 23%, approaching one zettabyte per year by the end of 2016 [1]–[3]. Traffic in metropolitan (metro) networks is growing twice as fast compared to long-haul traffic and, since 2014, accounts for more than half of the total. The faster growth is attributed to the increasingly significant role of content delivery, e.g., videos.

This traffic is predominantly handled by optical communication systems, and there are several options to scale these systems to higher data rates. The first option is to transmit more bits per symbol by using advanced modulation formats. This comes at the price of increased complexity and higher signal-to-noise power ratio requirements, which limit the reach of the communication system. The second option is to increase the symbol rate, which is limited by the bandwidth of the available electronics. The third and remaining option is to increase the number of parallel data channels. The common approach to realize this is to use multiple carriers at distinct frequencies. It is referred to as dense wavelength-division multiplexing (DWDM).

State-of-the-art DWDM systems transmit at a data rate of 100 Gbit/s per line card. Within each line card the data are modulated onto a single carrier. Driven by the continuous growth of data traffic, the research community searches for solutions to increase the data rate per line to 400 Gbit/s and 1 Tbit/s. The most promising approach is to group multiple carriers to a so-called super-channel and integrate them into a single line card. Within these super-channels, spectral guard intervals between the constituting sub-channels can be minimized. For line card manufacturers this means that the functionality of multiple line cards needs to be integrated into one. It is hence required that the individual components become smaller, cheaper, and more energy-efficient. In this thesis, we investigate using frequency combs as multi-carrier sources to replace a multitude of individual laser sources.

Optical frequency combs offer the advantage that only two parameters, the center frequency and the free spectral range, determine the full grid of frequencies. This removes the need for an individual frequency locking circuit for each carrier frequency, which would be needed in an array of individual lasers, thereby potentially reducing size and power consumption. The perfect and stable equidistance of the comb lines can be used to minimize spectral guard bands, as the relative drift between carrier frequencies is minimal.

In this thesis, three novel frequency comb sources are investigated, which can potentially be integrated into a line card, as multi-carrier generators for communication systems. We show that each of the investigated comb sources is capable of carrying data rates larger than 1 Tbit/s. First, gain-switched comb sources (GSCS), which offer easy tuning of line spacing and center frequency, are shown to be suitable for Tbit/s data transmission over several hundred kilometers. Second, quantum-dash mode-locked laser diodes (QD-MLLD) are tested. These sources feature a larger number of spectral lines and provide superior compactness, but suffer from a large optical linewidth. Two ways to overcome this limitation are demonstrated, leading to Tbit/s data transmission over 75 km. Third, very broadband frequency combs can be generated in nonlinear microresonators. These combs can span several telecommunication bands and are particularly well suited for multi-Tbit/s transmission systems.

Chapter 1 gives a general introduction into DWDM transmission schemes and the state of the art. We motivate the use of frequency comb sources for super-channel applications and discuss the required transmitter architecture. The comb sources investigated in this thesis will be introduced as well as the vision of a chip-scale Tbit/s transmitter module.

Chapter 2 summarizes the physical and technical background of optical frequency comb sources and the requirements that the comb lines need to fulfil in order to be used as carriers for coherent data transmission. The definition of an optical frequency comb used in this thesis will be explained and different methods of comb generation are summarized. Based on the requirements for high-speed data transmission, the relevant parameters of an optical frequency comb and the corresponding characterization techniques will be discussed.

Chapter 3 is dedicated to coherent Tbit/s data transmission using a GSCS. The comb generation technique and the experimental setup that are used to emulate a super-channel transmitter and receiver will be introduced. Enabled by the flexible line spacing of this comb source, several data transmission demonstrations with various symbol rates and modulation formats will be presented. Data rates of up to 1.9 Tbit/s could be transmitted over up to 300 km.

Chapter 4 explains the data transmission experiments performed with a QD-MLLD. This comb source excels by its compactness but also features high phase noise of the comb lines. Two ways to achieve coherent Tbit/s data transmission despite the drawback of high phase noise will be discussed. First, a feed-forward heterodyne phase noise reduction scheme can be implemented before modulating the data onto the carriers. A 1 Tbit/s data transmission over 75 km is demonstrated. Second, a self-homodyne detection scheme is implemented. Here the field of each comb line is split: One part serves as a carrier for the data, while the other part is transmitted to the receiver and used as a local oscillator. If both, the data stream and the unmodulated comb part arrive at the receiver with the same delay, their respective phase noise cancels when detecting coherently. This allows to transmit a 1.6 Tbit/s data stream over 75 km.

Chapter 5 discusses data transmission experiments using frequency comb sources that are generated with microresonators. Materials and resonator geometries as well as the comb formation dynamics will be introduced. It is found that the properties of the generated so-called Kerr combs strongly depend on the scheme for adjusting the parameters of the pump laser as well as on the dispersion profile of the resonator. The results of several data transmission experiments using different states of the Kerr comb will be discussed. The highest data rate is achieved with a soliton comb state, where a 19.7 Tbit/s data stream could be transmitted over 75 km.

Chapter 6 summarizes the main results, compares them to other comb-based coherent data transmission experiments, and gives an outlook on the integration potential and further improvements of the comb sources.

The work presented in this thesis is published in several journal publications and was presented at five international conferences; see page 167 ff. for a full list of publications.

Achievements of the present work

In this thesis, novel optical frequency comb sources, which can potentially be integrated into chip-scale Tbit/s transmitters, are investigated. For each comb source proof-of-principle coherent data transmission experiments are performed showing the potential for Tbit/s data transmission rates.

In the following, the main achievements presented in this thesis are highlighted. For all stated data rates, the bit-error-ratio is within the limits of standard forward-error correction techniques that operate with 20 % overhead.

Kerr frequency comb: Setup of a test-bench to generate and characterize microresonator-based Kerr frequency combs. First demonstration of coherent data transmission using microresonator-based frequency comb generation [J4], [C23]. Iterative improvements of comb and associated system setup [C31], [C23], [C15], [J1] lead to demonstration of data transmission at up to 19.7 Tbit/s over 75 km [C5].

Full C and L band data transmission: Demonstration of the first comb-based data transmission using a soliton Kerr comb, spanning the complete conventional and long wavelength transmission bands [C5]. The 19.7 Tbit/s data stream is transmitted over 75 km standard single mode fiber.

Stabilization of Kerr comb: Concept development, implementation and testing of a new feedback-stabilization scheme for Kerr frequency combs [J4], [C15]. Using the power generated by four-wave mixing as a feedback signal, the pump laser frequency could be effectively locked to a fixed position of the resonance.

Gain-switched comb source (GSCS): Tbit/s data transmission demonstrations using various line spacings, symbol rates, and modulation formats [J2]. Despite the limited bandwidth of the GSCS, up to 1.867 Tbit/s net aggregate data rate is achieved after transmission over 300 km. A maximum net spectral efficiency of 6.7 bit/s/Hz is shown requiring only 167 GHz bandwidth for a 1.1 Tbit/s super-channel.

Quantum-dash mode-locked laser diode (QD-MLLD): Design and setup of a test-bench for QD-MLLD, which are particularly sensitive to any kind of perturbations. Characterizations of mode-beating linewidth, optical linewidth and relative intensity noise are used to iteratively improve the test-bench.

Coherent data transmission with a QD-MLLD: For the first time, QD-MLLD are used for coherent data transmission. The large phase noise of the comb lines is reduced by using a feed-forward heterodyne phase-noise compensation scheme. A 1 Tbit/s data stream, based on QPSK modulation, is transmitted over a 75 km fiber span [C6]. Data transmission using 16QAM is achieved with the help of a self-homodyne system architecture. After 75 km transmission a 1.6 Tbit/s data stream could be measured [C7].

1 Introduction

In developed countries, access to the internet has become a commodity comparable to the supply of water or electricity. New services and applications, as on-demand content delivery and cloud access, keep driving the need for ever faster internet connections [1]–[3]. A saturation of this growth is not yet in sight as shown in Fig. 1.1, with data taken from [1] and [2]. The total annual data volume for metropolitan (metro) and long-haul networks is expected to surpass one zettabyte by the end of 2016. This corresponds to streaming about 2 million ultra-high definition videos simultaneously[1]. The compound annual growth rate (CAGR) is expected to be 23%. Mainly driven by video distribution, the metro sector will experience a stronger growth than long-haul links.

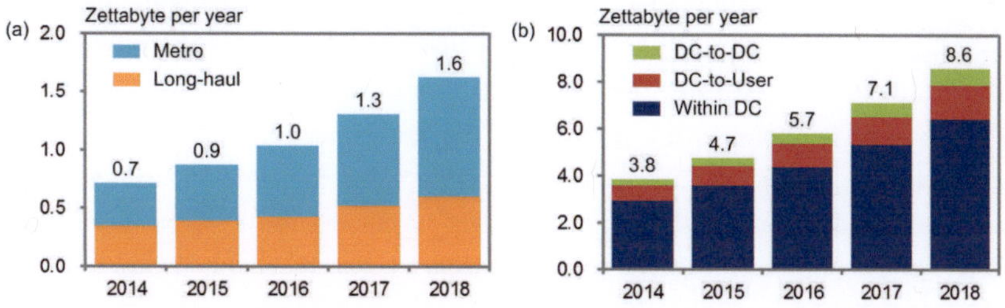

Fig. 1.1: Growth of global internet and data center traffic (DC) [1], [2]. For both cases the expected compound annual growth rate (CAGR) is 23%. One zettabyte corresponds to 10^{21} bytes. (a) Internet traffic through metro and long-haul fiber networks. (b) The total data center traffic is composed of traffic from one DC to another DC, between DC and user and within DC.

The total data center (DC) traffic is also projected to grow with a CAGR of 23% and to reach 8.6 ZB per year in 2018. This traffic comprises inter-DC traffic, i.e., data center interconnects (DCI), data-center-to-user traffic, and intra-DC traffic. Despite being the smallest fraction, the first one corresponds to the most strongly growing one with a CAGR of 29%, and the requirements here are quite similar to metro networks. The largest amount of traffic remains within the DC, although the share is expected to decrease slightly from 76.7% in 2013 to 74.5% in 2018 due to the strongly growing part of the inter-DC

[1] A Zettabyte per year corresponds to an average of 33 Tbit/s. An ultra-high definition video stream requires 18 Mbit/s [1].

connections. Note that the DC-to-user traffic from Fig. 1.1(b) is also included in the total internet data traffic shown in Fig. 1.1(a), as they share the same network.

The options to increase the data rates of the underlying optical communication systems are to increase the number of bits per symbol, the symbol rate, and the number of parallel data channels per fiber. Increasing the number of bits per symbol requires more distinct signal levels, and therefore poses higher requirements on the electrical and electro-optical components. Particularly the digital-to-analog (DAC) and analog-to-digital converters (ADC) need a higher resolution [4], [5], which is difficult to achieve at high sampling rates [6]. Furthermore, as the signal levels move closer together, the tolerance towards noise decreases, which ultimately limits the reach of a communication system [7]. Operation at higher symbol rates is challenging due to bandwidth limitations of the involved components and the reduced resolution of DAC and ADC. Additionally, the signal processing becomes more challenging, and eventually the electro-optical components might limit the bandwidth. Therefore, the preferred option for increasing the aggregate data rate is to increase the number of carrier frequencies. Nowadays, data signals from individual transceivers are multiplexed together to form a dense wavelength division multiplexing (DWDM) transmission system. Long-haul DWDM transmission systems are optimized for reach, which needs to be of the order of several thousand kilometers. For metro networks and DCI, however, cost, size and power consumption define the most important metrics, while the links are typically not longer than a couple of hundred kilometers.

State-of-the-art commercial transceivers typically transmit 100 Gbit/s or 200 Gbit/s [4], [8]–[12]. The transceivers rely on coherent data transmission, where data are encoded in both the amplitude and the phase of the optical carrier. This requires using an additional laser at the receiver that is used as a local oscillator (LO). The data are demodulated using digital signal processing (DSP). The transceivers operate at a symbol rate between 28 GBd and 32 GBd, use polarization-division multiplexing and a single carrier frequency. For a 100 Gbit/s, quadrature phase shift keying (QPSK) is used as modulation format, while 16-state quadrature amplitude modulation (16QAM) is employed for 200 Gbit/s. The first format transmits two bits per symbol, while the latter transports four bits per symbol. These formats are often called advanced modulation formats, and the corresponding transmitters and receivers are introduced in Appendix A. The research community is currently searching for the best option to increase the data rate of the transceivers to 400 Gbit/s, 1 Tbit/s and beyond. The most promising approach is to use multiple carrier frequencies as it removes the constraints of increasing the bandwidth of the electrical and electro-optical components. The challenge is then to integrate the functionality of multiple transceivers into a single module [13], [14].

Frequency combs are promising sources for such multi-wavelength transceivers as only two parameters, the center frequency and the free spectral range, need to be controlled to fully define the frequency grid rather than requiring control of each individual frequency in an array of independent lasers [15], [16]. Thus they offer a potential advantage in terms of size and power consumption. Furthermore, the remaining uncertainty of the grid spacing is removed, thus allowing to minimize spectral guard bands between the channels. This allows a more efficient use of the available bandwidth. Recently, the uncertainty of emission frequencies has been identified as the major impeding factor for compensation of nonlinear propagation impairments [17]–[19]. Using carriers derived from a frequency comb might therefore facilitate using higher launch powers into the transmission fiber and hence increase the reach of the communication system.

In the following sections, state-of-the-art technologies and standards for DWDM data transmission are reviewed and the concept of flexgrid networks and super-channels is introduced. Special attention is given to the optical sources, which might be replaced by frequency combs. Necessary changes to the system architecture are discussed and based on that, the vision of a chip-scale Tbit/s transmitter module is introduced.

1.1 Dense wavelength division multiplexing

In this section we review existing DWDM data transmission technology and standards. Emphasis lies on the optical sources for the DWDM data stream as well as on integration options, size and power consumption, which are a benchmark for the frequency comb sources that are investigated in this thesis.

1.1.1 Transmission windows and spectral grids

The operating window of DWDM systems is given by the wavelength region of minimum loss of the transmission medium, i.e., the standard single-mode fiber (SSMF). This minimum, as shown in Fig. 1.2, taken from [7], occurs around the wavelength of 1550 nm. The wavelength region is divided into several bands as indicated in the upper part of the figure. Most commonly used are the conventional (C) and long (L) wavelength bands since these two bands cover the wavelength region of minimum fiber attenuation and additionally benefit from the availability of the erbium-doped fiber amplifier (EDFA), which works for both bands.

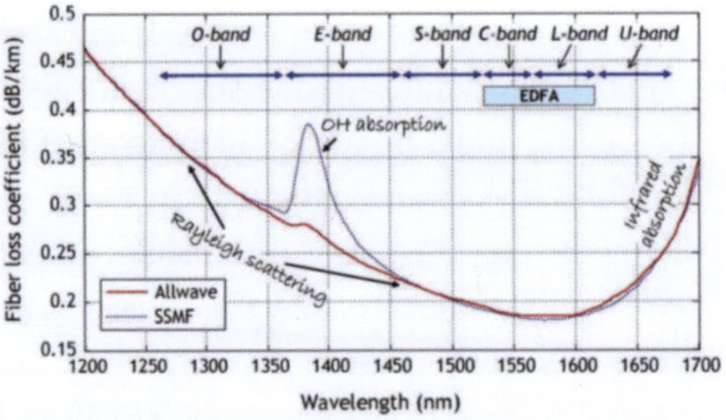

Fig. 1.2: Attenuation of optical fibers over wavelength [7]. The standard single-mode fiber (SSMF) is depicted in blue and a variant of it that is designed to remove the OH absorption peak (Allwave) is shown in red. The minimum loss occurs near the wavelength of 1550 nm, which is where the most commonly used C and L bands are located. These two bands coincide with the wavelength range that erbium-doped fiber amplifiers (EDFA) can amplify as indicated by the light blue box.

In this thesis comb sources operating in the C and L-Band are being investigated. In these bands, the spectral grid for DWDM is standardized by the ITU (recommendation G.694.1 [20]). The channel spacing is 12.5 GHz, 25 GHz, 50 GHz, or 100 GHz, and it is anchored to a central frequency of 193.1 THz. In commercial systems, the most commonly deployed spacing is 50 GHz. Considering the state-of-the-art data rate of 100 Gbit/s per wavelength channel, this amounts to a spectral efficiency (SE) of 2 bit/s/Hz. As the C-band can accommodate almost 100 wavelength channels, the total capacity in this case reaches nearly 10 Tbit/s. Scaling to higher data rates becomes challenging. A wavelength channel running at 400 Gbit/s would require an SE of 8 bit/s/Hz for the given 50 GHz grid. However, this needs a high signal-to-noise power ratio (SNR), which is only achievable over short distances [7]. The reason is that in order to increase the SNR a higher source power would be required. However, due to fiber nonlinearities the power to be launched into the transmission fiber must be kept below a certain threshold to avoid nonlinear distortions of the signal.

This means that future networks should support larger channel bandwidths to accommodate wavelength channels operating at 400 Gbit/s,

1 Tbit/s or 1.6 Tbit/s. However, network operators typically do not want to upgrade all line cards simultaneously, and so many line cards remain that operate at lower rates. Keeping a fixed channel spacing would hence result in a huge waste of bandwidth. This is depicted in Fig. 1.3(a) for the case of the most commonly used 50 GHz frequency grid.

A much more efficient use of the optical spectrum can be achieved by adapting the network to a flex grid architecture that is defined in the 2012 revision of the G.694.1 specification and is promoted by several big companies as Verizon [21], ADVA [22], Google [23], Infinera [14], and British Telecom [24], amongst others. The grid is based on a 6.25 GHz granularity for the center frequency and has slot widths of 12.5 GHz and integer multiples thereof. As shown in Fig. 1.3(b), the flex grid architecture can accommodate Tbit/s channels as well as other channel rates by allocating only as much bandwidth as necessary.

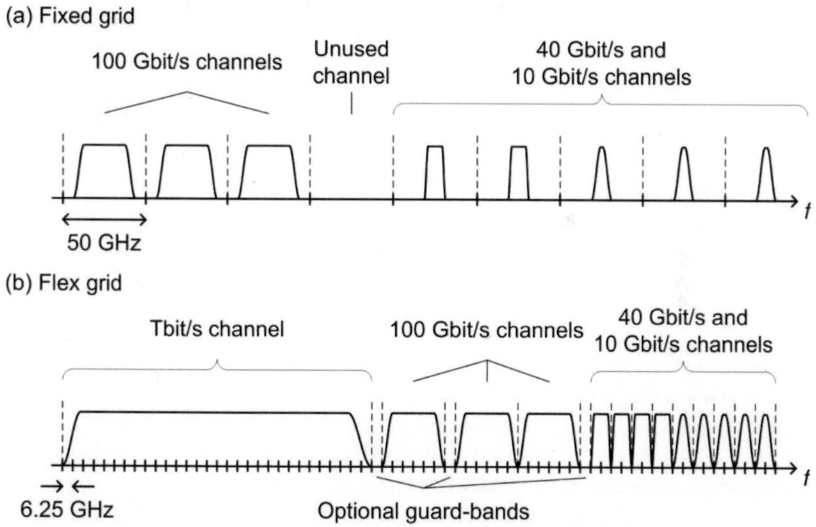

Fig. 1.3: Comparison of spectral grids according to G.694.1. (a) Conventional fixed 50 GHz grid commonly used in today's fiber networks. Each 50 GHz slot might be filled with a 100 Gbit/s, 40 Gbit/s, 10 Gbit/s channel, or left unused. (b) Proposed flex grid architecture with a center frequency granularity of 6.25 GHz and a minimum slot width of 12.5 GHz. This architecture could transport Tbit/s channels as well as the legacy data rates without wasting bandwidth.

In the previous paragraph, we already explained that there is a limit to the SE that can be achieved for a certain distance. This is discussed in detail in [7].

We therefore concluded that a higher bandwidth than the commonly used 50 GHz is required for channels operating at data rates on the order of Tbit/s. This bandwidth can be filled up efficiently by using optical super-channels [25], which are often discussed in conjunction with the flex grid network architecture [14], [22]–[24]. A good overview about Tbit/s super-channels as well as a review of experimental demonstrations data transmission can be found in [26].

A super-channel comprises an arbitrary number of sub-channels that are combined in a WDM scheme, and is characterized by a common routing from its origin to its destination. Within its allocated spectral window there is no requirement for a guard band, which would otherwise be needed to facilitate routing of the individual channels. There are many advantages of using super-channels. First, it allows to increase the per-channel data rate despite the bandwidth limitations of the electronics and opto-electronic components. Second, it allows to increase the spectral efficiency by reducing the amount of unused bandwidth, when being combined with the aforementioned flex grid network architecture. Third, the data rate can be changed on the fly by turning individual sub-channels on or off depending on the current traffic demand, which is interesting for applications in software-defined optical networks [26].

1.1.2 State of the art sources

Photonic integration is key to make multi-wavelength transceivers economically viable. The two most prominent integration platforms are silicon as well as III-V compounds semiconductors. A particularly challenging part is the integration of the required multi-wavelength source. The following sections discuss state-of-the-art sources as well as research towards multi-wavelength sources that are not based on comb generation.

Integrable tunable laser assembly (ITLA)

For a DWDM system, the most commonly used light source is an external cavity laser (ECL) that is specified and packaged according to the multi-source agreement (MSA) for integrable tunable laser assemblies (ITLA)[2]. The MSA defines physical dimensions, interfaces and minimum performance requirements for the laser assembly. The dimension of the whole module, including control electronics with pre-defined interfaces, is 74 mm × 30.5 mm

[2] http://www.oiforum.com/public/documents/OIF-ITLA-MSA-01.2.pdf

× 10.5 mm. These lasers are typically tunable over the C or the L band, feature a fiber coupled output power of the order of 16 dBm, and offer low relative intensity noise (RIN), < -140 dB Hz^{-1}, and a linewidth < 1 MHz. They are particularly well suited for long-haul and metro transmission distances, exploiting advanced modulation formats at high symbol rates. The power consumption is of the order of several watts, a maximum of 6.6 W is specified in the implementation agreement. Typically, a power consumption of 3 W is found in the data sheets[3].

With the given size and power consumption of the ITLA, it is unlikely that several tens of these sources are combined for high-speed super-channel data transmission. Some improvement is found by adopting the newer implementation agreement for a micro-ITLA, which reduces the size to a maximum of 65 mm × 20 mm × 7.5 mm with a maximum power consumption of 5 W. These sources will likely suit a first generation of Tbit/s super-channels with a maximum of 5 to 10 sub-channels. But radically new approaches are required to achieve a significant reduction of size and power consumption.

Vertical cavity surface emitting lasers (VCSEL)

Compact and very efficient laser sources are the so-called vertical cavity surface emitting lasers (VCSEL) [27], [28]. Such lasers can typically operate at currents of the order of 20 mA, consume 40 mW of power and deliver a few mW of optical output. Furthermore, due to the vertical emission they allow for wafer-level testing, thereby significantly reducing cost. Single-mode devices emitting in the C-band have been developed [29]. Recent work also shows that VCSEL can potentially be used for coherent data transmission [30]. However, the linewidth of 5 MHz limits their use for higher-order modulation formats. Apart from this limitation, they are not very well suited for integrated WDM systems: The emission frequency strongly depends on current and temperature. A shift of hundreds of GHz per mA and tens of GHz per K means that a very low-noise current source and a stable thermo-electric cooler (TEC) would be required to stabilize each VCSEL to the desired grid frequency.

[3] http://www.emcore.com/wp-content/uploads/EMCORE-ITLA-High-Data-Rate-Application-Note.pdf

Laser array on InP wafers

A good and successful example of monolithic integration of a Tbit/s super-channel is provided by the company Infinera [31]–[33]. Starting from InP wafers, III-V compound semiconductors are used to monolithically integrate an array of distributed feedback (DFB) lasers with modulators, multiplexers, amplifiers, and receivers. Infinera's current commercial product implements 10 channels operating at 14.3 GBd using quadrature phase shift keying together with polarization division multiplexing, and achieves a super-channel with a net data rate of 500 Gbit/s [33]. Scaling of the photonic integrated circuit (PIC) to 40 channels has been demonstrated [34]. However, it remains to be shown that simultaneous operation of all the lasers and data channels is feasible: The emission wavelength of DFB lasers suffers from the same strong temperature and current dependence as VCSEL. The author of [35] argues that such a number of parallel DFB lasers is not practical in view of power consumption and cost.

The size and power consumption of a laser array can be estimated from the values for a single laser chip as offered for example by Emcore[4]. The lateral chip dimensions are 300 μm by 254 μm. The specified typical operating voltage is 1.3 V. Setting the laser current to 100 mA leads to an optical output power of 36 mW (15.6 dBm). A laser driver[5] for such a device would consume approximately 400 mW and the temperature controller[6] needed for wavelength stabilization adds 1.75 W to the total power consumption. In an array of such lasers, size and power consumption scale linearly. Additional overhead might be needed to avoid thermal crosstalk.

Hybrid silicon III-V lasers

Silicon photonics is a promising integration platform, however, silicon lacks the ability to generate light. A hybrid integration approach using III-V material on silicon is a possible way to circumvent this deficiency. In this case, the optical gain is provided by the III-V material, and the silicon waveguides are part of the laser cavity. A wafer bonding approach has been demonstrated in [36], which allows to build a DFB laser with 3.6 MHz linewidth [37]. The technique uses evanescent coupling between the III-V gain material and the

[4] http://www.emcore.com/g1033-035-1550-nm-gpon-laser-diode-chip/
[5] http://www.teamwavelength.com/products/product.php?part=147&view=specs#tabs
[6] http://www.maximintegrated.com/en/products/power/switching-regulators/MAX8520.html

silicon waveguides. Alternatively, the authors of [38] use a flip-chip approach to bond a reflective semiconductor optical amplifier onto a silicon grating coupler in order to build a laser. The silicon waveguides become part of the laser cavity and can incorporate different filtering options to achieve wavelength tunability or even multi-wavelength lasers, e.g., arrayed waveguide grating [39] and ring resonators [40]. Recently, mode-locked lasers have been realized in such a hybrid III-V approach on silicon with a repetition rate of 4.7 GHz [41]. The optical linewidth of this comb source was investigated in [42], where injection locking is proposed to reduce the linewidth from several MHz down to the linewidth of the injection source. However, this scheme increases complexity of the system, and it reduces the number of emitted comb lines. While these are exciting fields of research, up to now, there has been no demonstration of a multi-wavelength source with a reasonably large number of lines, an appropriately large power per line as well as low linewidth and RIN.

1.2 Data transmission with frequency combs

Using optical frequency combs as multi-carrier sources for super-channel data transmission has several advantages such as a reduced number of required laser sources and wavelength locking circuits. Potentially, this reduces size and power consumption and increases the attainable spectral efficiency leveraging the inherently equidistant comb line spacing [15], [16]. On the other hand, such systems require either a demultiplexer to spatially separate the comb lines, or an array of wavelength selective modulators to facilitate modulating each comb line individually with data.

The latter approach uses for example ring resonator modulators [43] and is also likely to become the technology-of-choice for the first commercial comb-based data transmission system[7]. This system is targeted for data center applications, and is particularly compact and energy-efficient supporting a reach of up to 2 km[8].

This thesis investigates systems relying on demultiplexing the lines of the comb, and individually modulating the resulting carriers. This allows the use of advanced modulation formats in order to maximize spectral efficiency. The corresponding system architecture will be introduced in the following section.

---

[7] http://ranovus.com/
[8] http://openopticsmsa.org/

1.2.1 System architecture

The basic architecture for a comb-based super-channel transmitter is depicted in Fig. 1.4. In this example the comb consists of 10 lines represented by the colored spectrum on the left. A wavelength demultiplexer (DEMUX) separates the comb lines spectrally such that they can be individually modulated, here with a data rate of 100 Gbit/s. The data channels are then multiplexed (MUX) to form the Tbit/s super-channel. Such a dense channel spacing as indicated by the colored spectrum on the right required minimal interference from channel crosstalk. Nyquist-WDM or orthogonal frequency-division multiplexing (OFDM) [44] can achieve this goal.

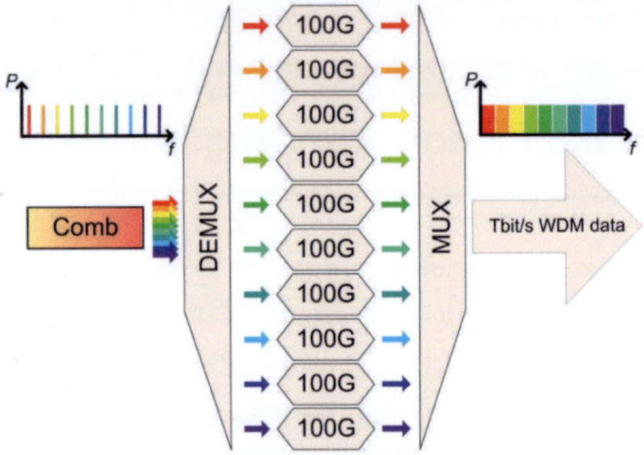

Fig. 1.4: Principle of a comb-based super-channel transmitter. The comb lines are demultiplexed, individually modulated and multiplexed to form the super-channel. A Tbit/s data stream can for example be achieved by using 10 comb lines and modulating each with a 100 Gbit/s signal.

There exist a variety of options to realize the DEMUX and MUX. For example ring resonator-based add/drop filters [45], [46], echelle gratings [47], arrayed waveguide gratings [48], or cascaded interleavers based on Mach-Zehnder interferometers [49]. For the experiments described in this thesis the comb lines are only dis-interleaved into even and odd carriers rather than to fully demultiplex the comb. This is a common technique to emulate WDM systems or super-channel transmitters in a laboratory environment requiring only one pair of data sources and modulators.

1.2.2 Comb sources investigated in this thesis

Frequency combs have already shown their high potential as multi-wavelength sources for transmitting Tbit/s data streams with high spectral efficiency [50]–[53]. However, the comb generators that were used for these demonstrations are too bulky and too expensive to be a realistic alternative for systems currently being deployed. In this thesis three different comb sources are investigated. All of them are suitable for carrying Tbit/s data streams. They feature promising options for dense integration. This section gives a short introduction to the operation principle of each comb source and discusses the individual challenges that will be addressed in detail in Chapters 3 to 5.

In Fig. 1.5, the basic setup for each of the three comb sources can be seen: The gain-switched comb source (GSCS), the mode-locked laser diode (MLLD), and the Kerr comb.

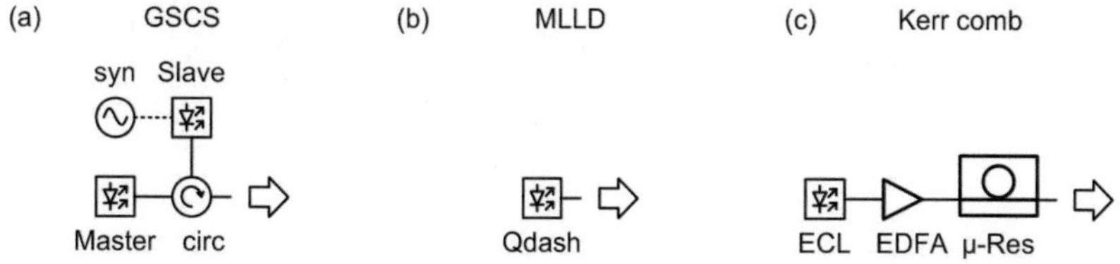

Fig. 1.5: Investigated comb sources. (a) The so-called gain-switched comb source (GSCS) consists of a slave laser that is being modulated by a synthesizer (syn) thereby generating comb lines with a spacing that corresponds to the modulating frequency. The master laser injects light into the slave cavity via a circulator (circ) to improve linewidth and relative intensity noise of the comb lines. (b) The mode-locked laser diode (MLLD) is a Fabry-Perot-type laser with a gain material that comprises quantum-dashes (Qdash) in the gain material. Multiple longitudinal modes oscillate simultaneously, thus forming a frequency comb. Mode-locking occurs via third-order optical nonlinear interaction inside the cavity. (c) The so-called Kerr comb is generated by cascaded four-wave mixing processes induced by strong pumping of a highly-nonlinear microresonator (μ-Res).

Gain-switched comb source

The GSCS, Fig. 1.5(a), was developed at Dublin City University and is currently being commercialized by the start-up company Pilot Photonics. It comprises a synthesizer (syn) that modulates the current of a slave laser, thereby generating short pulses with a repetition rate corresponding to the modulation frequency [54]–[56]. The master laser injects light into the slave cavity, which correlates the phase of the slave laser oscillation between the individual pulses [57]. Thus the corresponding spectrum is a frequency comb with comb lines that feature low linewidth and low RIN [58]–[60].

To understand the working principle of gain-switching, recall that the step response of a laser exhibits relaxation oscillations [56]. For generating a train of short pulses the laser is repetitively switched on for such a short time that only the first spike of the relaxation oscillation is generated [56], [61]. However, if the pulses develop out of spontaneous emission, there is no coherence between the pulses. Hence, they exhibit a huge time jitter, which results in a broadband spectrum without distinguishable comb lines [58]. This problem can be solved by injection locking. To this end, CW light from a second laser is injected into the modulated laser [57]. The presence of the injected photons causes stimulated emission during the pulses and establishes coherence between the pulses. This reduces the timing jitter, and the corresponding spectrum becomes a stable frequency comb [58].

The advantages of this comb source are the electrically tunable free spectral range and center frequency, the spectral flatness, and the stable operation. However, the 3 dB bandwidth of the comb is only about 200 GHz and thus a very high spectral efficiency is required to use these sources for Tbit/s data transmission. In Chapter 3 it is shown that GSCS are indeed suited as carrier for Tbit/s data streams [J2]. A maximum of 1.87 Tbit/s was transmitted over up to 300 km. The spectral efficiency in this case amounts to 6.2 bit/s/Hz.

Mode-locked laser diode

The MLLD [62], [63], Fig. 1.5(b), consists of a single laser chip. For operation it suffices to apply a direct current, to employ a temperature controller, and to use a proper chip-to-fiber coupling. The laser is a Fabry-Perot type with cleaved facets acting as front and back-side mirrors. The inhomogeneously broadened gain spectrum provided by the quantum-dash (Qdash) gain material allows multiple longitudinal modes to oscillate simultaneously. Mode-locking that can be observed for selected operating points, is attributed to third-order optical nonlinear interactions inside the cavity [63]–[65]. The laser chips are developed, designed, and fabricated at III-V Lab and at the Centre National de la Recherche Scientifique (CNRS) at Marcoussis, France.

The advantages of this comb source are its superior compactness, low power consumption, and broad comb spectrum with a 3 dB bandwidth of about 1.6 THz. However, the comb lines suffer from large phase noise that is prohibitive for advanced modulation formats, which, however, are required for a high spectral efficiency. Chapter 4 discusses the operation and characterization of the lasers and presents two options for transmitting a Tbit/s data stream despite the phase noise issue [C6], [C7].

Phase noise compensation: A feed-forward phase noise compensation scheme is used, in which compensating phase noise improves the quality of selected comb lines. Transmission of a 1 Tbit/s data stream over 75 km is demonstrated with the help of this technique [C6].

Self-homodyne architecture: This option uses a self-homodyne system architecture. In this case the field of each comb line is split: One part serves as a data carrier, while the other part is transmitted to the receiver and serves as a local oscillator. If the modulated part and the unmodulated part of the comb lines arrive at the receiver with the same delay then their respective phase noise cancels when detecting coherently. With this technique we could demonstrate transmission of 1.6 Tbit/s over 75 km [C7].

Kerr comb

The Kerr comb, Fig. 1.5(c), exploits cascaded four-wave mixing by strongly pumping a highly nonlinear microresonator [66]. The details of the comb generation dynamics and the underlying physics are introduced at the beginning of Chapter 1 and in Appendix A. The resonators used for the data transmission experiments as described in Chapter 5 were designed, developed, and fabricated at École Polytechnique Fédérale de Lausanne and at the Femto-ST in Besançon. The outstanding feature of the Kerr comb is the huge bandwidth of the generated comb that can even span a whole octave [67]. Basically, such a comb source requires nothing more than a strong CW laser and a microresonator. However, as a rule, the resulting combs are unsuitable for data transmission [C31] because of the formation of many overlapping sub-combs. These sub-combs share the same line spacing δ, but they can be slightly offset from each other [68]. As a result, the complete comb spectrum consists of densely packed multiplet lines, which leads to strong amplitude and phase noise. There are a number of measures to improve the situation, as will be explained in the next three paragraphs.

Primary comb: The initial comb, i.e., the comb before the sub-combs are generated, consists of a small number of primary comb lines that are typically spaced by an integer multiple m of the free spectral range (FSR) of the micro-resonator. For later reference this spacing is denoted as Δ. These lines are quite robust against perturbations of the pump parameters and inherit the noise properties of the pump laser. They are hence well suited as carriers for coherent communications [J1] as discussed in Chapter 5.2.2. However, the power of the lines strongly drops with spectral distance from the pump.

Δ-δ-matching: It has been shown in [68] that it is also possible to influence the offset of the sub-combs by fine-tuning of the pump parameters in order to achieve $\Delta - m\delta = 0$. This technique is referred to as $\Delta - \delta$-matching. In this case the comb lines can be used for coherent data transmission [J4], which is discussed in Chapter 5.2.3. However, the drawback of this method is the uneven spectral envelope, which is attributed to constructive and destructive interference between the contributions of each sub-comb to an individual comb line.

Soliton comb: Additional dispersion engineering [69], [70] along with an advanced pumping scheme allows to generate solitons inside the microresonator [71]. These solitons rely on a double balance between anomalous dispersion and self-phase modulation on the one hand and between cavity loss and pump-induced parametric gain on the other hand. Their spectrum features a large number of comb lines with a spectral envelope that resembles a sech^2-function. Chapter 5.2.4 discusses the first coherent data transmission experiment performed with such a soliton Kerr comb, where 19.7 Tbit/s were transmitted over 75 km [C5]. This corresponds to the highest data rate achieved to date with a chip-scale frequency comb source.

1.2.3 The vision: Chip-scale Tbit/s transmitter

In this section, the vision of a chip-scale Tbit/s transmitter module is described using the example of a Kerr frequency comb source. The explanations and Fig. 1.6 are taken from [J4]. Naturally, the Kerr comb source could be replaced by the other two comb generators, which were previously introduced.

The vision of a future chip-scale Tbit/s transmitter is illustrated in Fig. 1.6(a). A Kerr frequency comb is generated by exploiting multi-stage four-wave mixing (FWM) in a high-Q Kerr-nonlinear microresonator pumped with a strong continuous wave laser [66], [72]. The envisaged transmitter consists of a multi-chip assembly, where single-mode photonic wire bonds [73] connect the individual chips. In contrast to monolithic integration, this hybrid approach combines the advantages of different photonic integration platforms: For the optical pump, III–V semiconductors can be used [31], while the high-Q ring resonator for Kerr comb generation could be fabricated using low-loss silicon-nitride (SiN) waveguides [74]. The optical carriers are separated and individually modulated on a transmitter chip, where large-scale silicon photonic integration lends itself to particularly compact and energy-efficient MUX and DEMUX filters [75] and IQ modulators [76], [77].

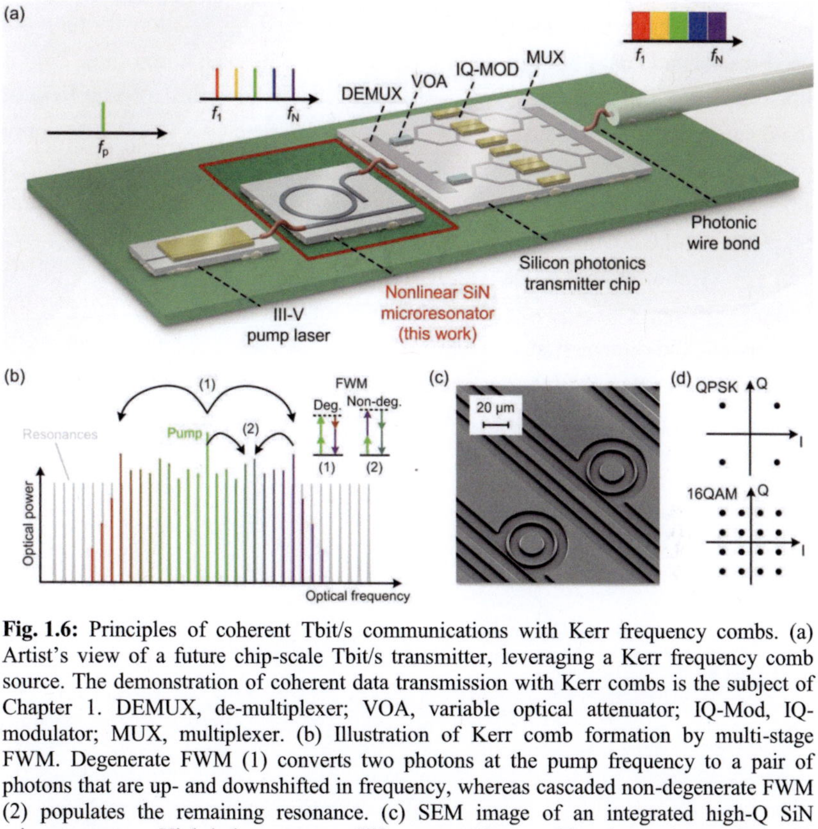

Fig. 1.6: Principles of coherent Tbit/s communications with Kerr frequency combs. (a) Artist's view of a future chip-scale Tbit/s transmitter, leveraging a Kerr frequency comb source. The demonstration of coherent data transmission with Kerr combs is the subject of Chapter 1. DEMUX, de-multiplexer; VOA, variable optical attenuator; IQ-Mod, IQ-modulator; MUX, multiplexer. (b) Illustration of Kerr comb formation by multi-stage FWM. Degenerate FWM (1) converts two photons at the pump frequency to a pair of photons that are up- and downshifted in frequency, whereas cascaded non-degenerate FWM (2) populates the remaining resonance. (c) SEM image of an integrated high-Q SiN microresonator. High-index-contrast SiN waveguides enable dense integration. (d) Constellation diagrams of QPSK and 16QAM signals, where information is encoded both in the amplitude and the phase of the optical carrier, which can be represented by the in-phase (I, horizontal axis) and quadrature (Q, vertical axis) components of the complex electrical field amplitude.

Fig. 1.6(b) illustrates the basic principle of Kerr comb generation. Pump energy is transferred to the comb lines by two processes: Degenerate FWM, indicated by (1) in Fig. 1.6(b), leads to the formation of two side modes by converting two pump photons into a pair of photons that are up- and downshifted in frequency. The magnitude of the frequency shift is determined by the pump power and by the cavity dispersion [68]; a multitude of cascaded non-degenerate FWM processes, indicated by (2) in Fig. 1.6(b), fully populate the remaining resonances. Kerr comb generators can be extremely compact, see Fig. 1.6(c), which shows a scanning electron microscope (SEM) image of a planar integrated SiN microresonator.

To maximize the spectral efficiency, advanced modulation formats are used that encode data both in amplitude and phase of each comb line. This is illustrated in Fig. 1.6(d) for QPSK and 16QAM. The complex electric field of the data signal is visualized by its in-phase (I, horizontal axis) and quadrature component (Q, vertical axis) in the complex plane. For QPSK, the I- and Q-components can assume two distinct values, leading to four signal states (symbols) in the complex plane. This represents an information content of two bits per symbol. Similarly, a 16QAM symbol can assume 16 states in the complex plane, corresponding to four bits of information per symbol. A densely packed optical signal spectrum, as schematically depicted in the upper right corner of Fig. 1.6(a), can be achieved by using sinc-shaped Nyquist pulses having rectangular power spectra [44].

2 Fundamentals

The main purpose of this thesis is to investigate the use of optical frequency combs as a multi-carrier generator for coherent data transmission applications at Tbit/s data rates. This chapter provides background information for the experiments that were carried out during this thesis. Section 2.1 presents the basics of optical frequency combs. We describe the different methods to generate frequency combs and relate them to the techniques used in this thesis that are explained in chapter 1.2.2. Section 2.2 introduces the parameters of the optical frequency comb that are relevant for coherent data transmission. Fundamentals of coherent data transmission and reception will be introduced in Appendix A.

2.1 Frequency combs

2.1.1 Definition

A frequency comb, as the term is used throughout this thesis, is a set of narrow-band lines, equidistantly spaced over frequency as depicted in Fig. 2.1.

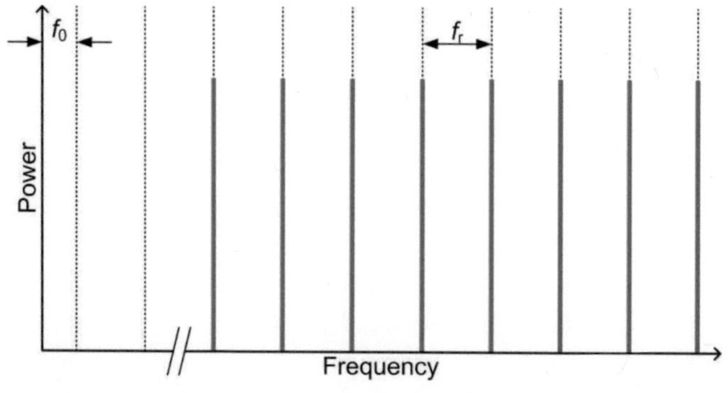

Fig. 2.1: Frequency comb. A frequency comb is a set of narrow-band equidistantly spaced lines. The frequencies are defined by two parameters, the free spectral range f_r and the offset frequency f_0.

All frequencies f_m of the comb are fully determined by only two parameters, the free spectral range f_r and the offset frequency f_0. The latter is the frequency of the hypothetical zeroth comb line,

$$f_m = f_0 + m \cdot f_r . \tag{1.1}$$

For the case of optical communications the definition of frequencies for a wavelength-division multiplexing (WDM) system is given in terms of a center frequency, i.e., 193.1 THz [20]. For a dense WDM system the channel spacings are multiples of 12.5 GHz with the multiplier being 1, 2, 4 or 8.

Note that certain communities only speak of a frequency comb if both f_0 and f_r are precisely known and controlled, which requires very broadband combs, spanning more than an octave [78]. However, in the context of data transmission, the term frequency comb is used for a set of a few equidistant lines.

2.1.2 Methods of comb generation

Mode-locked lasers

Historically, optical frequency combs were developed for high precision absolute frequency measurements. The idea behind this is to link optical frequencies to a microwave frequency that can be precisely measured. For this purpose, broadband frequency combs spanning more than an octave have been developed. The line spacing of these combs is typically in the MHz range. It is then possible to precisely measure and control f_0 and f_r. This can be achieved with mode-locked lasers (MLL) and was awarded with the Nobel prize in 2005 [79], [80]. There is a number of review papers and books discussing the development of mode-locked lasers and giving detailed explanations of their working principle, see for example [78], [81]–[83].

In this section, the basic working principle of a MLL will be explained briefly with a focus on schemes that are relevant for optical data transmission. This means that the free spectral range needs to be at least of the order of several GHz. Furthermore, the discussion focuses on lasers emitting in the wavelength range of the C-band according to Fig. 1.2. Additional requirements for the frequency comb will be discussed in more detail in Section 2.2.

An MLL is a laser that emits multiple longitudinal modes, which are locked in a way that the superposition of the fields with different equidistant frequencies forms a train of periodic pulses in the time domain, see Fig. 2.2. In the figure the pulsed output comprises four longitudinal modes (Field 1…4), which constructively interfere at times corresponding to integer multiples of the cavity round trip time τ_r. The spectra of the four fields and the resulting pulse train are depicted on the right hand side. They correspond to equidistant lines with a free spectral range (FSR) equal to $1/\tau_r$.

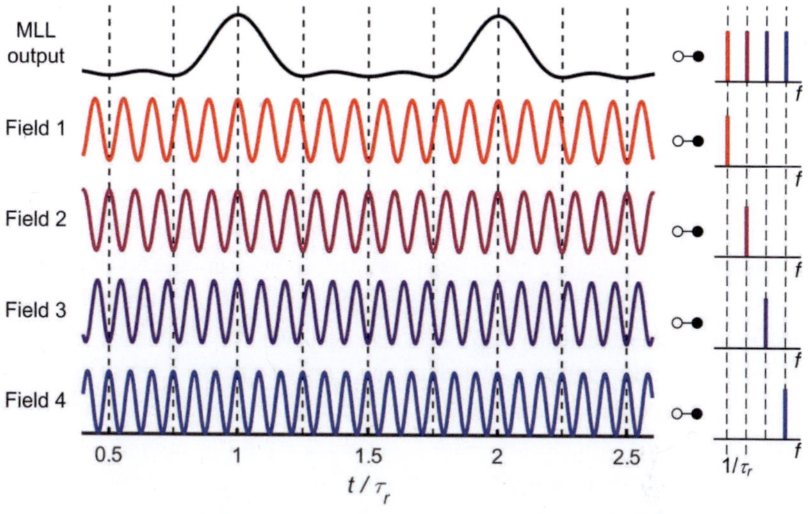

Fig. 2.2: Field of a mode-locked laser (MLL) in time and frequency domain. In this simplified example the MLL output comprises four longitudinal modes termed field 1…4 of the MLL. At each integer multiple of the cavity round trip time τ_r, the time signals interfere constructively resulting in a strong output pulse. The respective Fourier transforms are depicted on the right hand side.

There are different techniques to achieve mode locking. One is to modulate the cavity losses at a frequency that corresponds to the free spectral range of the cavity. This technique is referred to as active mode locking. The generation of pulses can be understood by considering that the modulation of the cavity losses periodically turns the laser on and off at a rate corresponding to the cavity round trip time. Another possibility is to use a saturable absorber or a Kerr-lens in combination with an aperture. In both cases, the cavity losses are modulated by the pulse itself, i.e., the pulse experiences less losses at its peak than at its wings. This technique is referred to as passive mode locking. Additionally, there are hybrid approaches that use a combination of a saturable absorber and active modulation of the cavity losses.

The best performing MLL are built with large discrete components. Hence they feature a large cavity round trip time, and the mode spacing is in the MHz range. One way to increase the mode spacing is harmonic mode-locking [84]. In this case, multiple pulses are circulating simultaneously in the cavity. However, the problem of having a large size of the laser is not solved but rather increased due to the additional arrangements needed to achieve stable harmonic mode-locking.

Several attempts have been made to reduce the cavity length in order to achieve fundamental mode-locking with a mode spacing in the GHz range. In a free-space setup using a semiconductor saturable absorption mirror (SESAM) [85], the cavity length could be reduced down to 1.26 mm, yielding a repetition rate of 100 GHz [86]. However, the generated frequency comb consists of very few comb lines only. The number of comb lines can be increased by spectral broadening as described later in this section, but the associated overhead diminishes the advantage of having a compact comb source. Another option is to increase the number of lines by a factor of n by time interleaving the optical pulses from an MLL [87]. That means the pulses are split into n paths and recombined with respective delays corresponding to $\tau_r/n ... (n-1)\tau_r/n$. Although this can be integrated on a chip and thus adds less overhead as compared to the spectral broadening, it suffers from a couple of drawbacks. First this technique does not extend the bandwidth of the comb but rather adds the lines in between the existing ones, and second it needs to be adapted precisely to the respective comb source. This leaves little room for fabrication tolerances.

Another compact MLL was achieved by sandwiching an erbium-doped fiber between two single-mode fibers with high-reflectivity (HR) coatings on their facets and a carbon nanotube based saturable absorber on one of these HR coatings [88]. The authors demonstrate a free spectral range of up to 19.45 GHz with this approach. However, it remains unclear whether this comb features the stability and low-noise behavior necessary for data transmission.

The most promising option to reduce the size of an MLL is to use integrated photonic circuits. Monolithically integrated MLL using III-V semiconductors [89], [90] have been investigated as well as hybrid approaches where a III-V gain material is combined with silicon-on-insulator waveguides [91]. The compact size of monolithically integrated MLL makes them interesting for the envisaged multi-Tbit/s transmitter introduced in Section 1.2.3. As mentioned in Section 1.2.2 we investigate in chapter 1, a particularly interesting implementation of an integrated MLL, the quantum-dash mode-locked laser diode (MLLD) [62], [63].

Modulator-based approaches

Frequency combs with exceptional flexibility in terms of center frequency and free spectral range can be generated by sinusoidal modulation of a continuous wave (CW) laser. The frequency of the CW laser determines the center frequency of the comb, and the frequency of the sinusoid determines the free spectral range. A summary of different approaches for modulator-based comb generation is given in [92]. Modulator-based comb sources offer the possibility to shape the spectral envelope of the comb to achieve a flat-top spectrum. This has been demonstrated with a phase modulator driven by two sinusoidal tones [93], with cascaded amplitude and phase modulation [94], with dual-drive Mach-Zehnder modulators [95], [96], with an in-phase quadrature phase (IQ) modulator [97], and with cascaded phase modulators [15]. However, as the voltage U_{π} that is required to achieve a π phase shift, of conventional modulators is quite large, this scheme needs very high-power electrical drivers. This drawback can be mitigated to some extent by using integrated modulators, e.g., InP [98], [99], or silicon organic hybrid modulators [J5]. The latter approach is demonstrated to be suitable for data transmission at Tbit/s rates. In Chapter 6.1 it is compared to the other comb generation techniques investigated in Chapters 1 to 1.

As introduced in Section 1.2.2 it is also possible to directly modulate the current of a laser [54]–[56]. The properties of these combs can be greatly improved by injection locking [58], [59], [100]. In Chapter 1 we discuss the use of these so-called gain-switched comb sources (GSCS) for Tbit/s coherent optical data links [J2].

Modulator-based comb sources typically deliver a few comb lines only. In order to increase the number of lines, it is possible to cascade multiple modulators [101], [102]. Alternatively, the modulator can be put into a loop. This technique is referred to as a recirculating frequency shifter (RFS). In [103], an IQ modulator, driven as frequency shifter, i.e., operating as a single side-band modulator driven by a sine wave, was placed in a loop together with a filter and a Raman amplifier to generate 50 lines spaced by 12.5 GHz within a flatness of 2.5 dB. Even more lines were generated with cascaded phase modulators in an RFS. In [104] an RFS was used to generate 113 lines within a 5 dB flatness and with an FSR of 25 GHz. An integrated version of an RFS-based comb generator was shown in [105] using the InP platform, achieving 6 lines with 5 dB flatness and with 10 GHz spacing.

As a summary for modulator-based comb generation, it can be stated that there is a trade-off between the number of lines, flatness, and the setup's complexity. A further disadvantage is the need for power-hungry high frequency electronics. However, it is the most flexible approach in terms of freely adjustable center frequency and FSR. Furthermore, it is a stable and easy-to-use comb source, delivering high-quality carriers that are very well suited for data transmission applications.

Comb generation based on third-order optical nonlinearity

In this section the possibilities of generating frequency combs using third-order optical nonlinear interactions will be discussed. In this context new frequencies are generated via self-phase modulation (SPM), cross-phase modulation (XPM) and four-wave mixing (FWM). For details on the underlying physics, the interested reader is referred to [106]–[108].

Parametric techniques are generally used to create very broadband frequency combs. Launching two lines that are offset in frequency into a highly nonlinear fiber (HNLF) can induce strong nonlinear effects creating a multitude of new lines spaced by the offset of the initial two lines. This requires high launch powers and a careful engineering of the chromatic dispersion of the HNLF. It is also crucial that the initial lines are phase locked, as uncorrelated phase noise multiplies towards the outer spectral lines [109]. For this purpose a frequency comb with only a few lines can be generated using, e.g., a modulator-based technique. From this comb, two lines can be used for injection locking two separate lasers, which provides a phase sensitive amplification to generate two phase-locked high power tones to be launched into the HNLF. It is also possible to directly use the few-line comb as seed for the nonlinear broadening. By carefully arranging multiple nonlinear mixing and regenerative stages for pulse shaping, a comb with more than 1500 coherent lines within 2 dB spectral flatness and a spacing of 10 GHz can be generated [110].

Such spectral broadening techniques have also been used to increase the bandwidth of MLL. For example in [111], a 50 GHz MLL is broadened in a HNLF. Using a dynamic gain equalizer the broadened comb is flattened to exhibit 37 lines within 3.5 dB flatness. The spectral flatness is limited by the modulations in the comb's center, which are typical for spectra broadened by SPM. A different way to overcome these modulations is to spectrally slice the original comb lines at the center of the broadened spectrum [112]. Here a total of 325 comb lines is generated and spectrally flattened.

Kerr frequency combs as introduced in Section 1.2.2 exploit nonlinear effects in a microresonator [66] and provide a path towards a hybrid integration of such broadband frequency combs, Fig. 1.6. The application of such combs for data transmission is discussed in detail in Chapter 1.

2.2 Comb parameters, requirements and measurement

Using frequency combs for data transmission poses unique requirements on the parameters of the comb such as line spacing, power per line as well as intensity and phase noise of the comb lines. In this section, these properties and associated characterization techniques are explained. We discuss how each of the parameters might limit the data rate that can be obtained with a given comb source.

2.2.1 Free spectral range, power and intensity noise

The best starting point to characterize a frequency comb is to measure its spectrum using an optical spectrum analyzer (OSA). Fig. 2.3 shows a sketch of such a spectrum and illustrates the characteristics that can be extracted from such a measurement. These are line spacing or FSR, power of each comb line, flatness (variation of line power for a certain number of comb lines), and optical carrier-to-noise power ratio (OCNR). The FSR directly gives an upper limit to the symbol rate with which each line can be modulated, while power and OCNR constrain the choice of modulation formats.

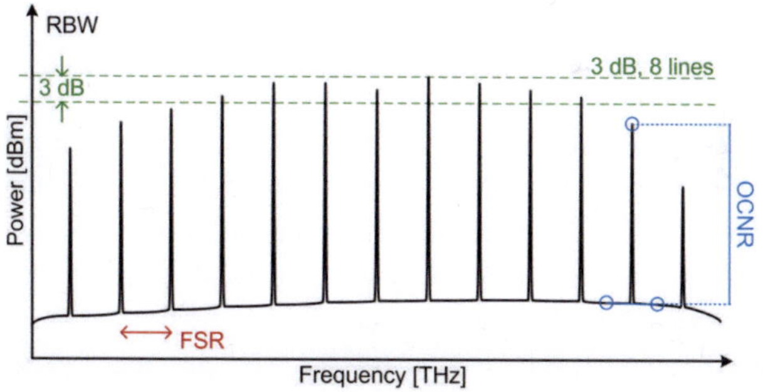

Fig. 2.3: Sketch of a spectrum illustrating the main characteristics of a frequency comb: Comb flatness (variation of line power) for a certain line number, free spectral range (FSR), and optical carrier-to-noise power ratio (OCNR) for the given resolution bandwidth (RBW).

Free spectral range

For data transmission with frequency combs, an FSR in the range between 10 GHz and 100 GHz is desirable. Higher FSR have the advantage that the comb lines can be separated with available DEMUX filters, which is required for the system architecture introduced in Section 1.2.1. However, for too large an FSR, the advantage of using a comb source compared to employing individual lasers diminishes because only very large symbol rates would lead to a high spectral efficiency. On the other hand, for lower FSR it becomes increasingly difficult to separate the comb lines for individual data modulation. In addition, the number of comb lines required to achieve a reasonably high data rate becomes large, and so does the number of data sources and modulators. However, by choosing an FSR in the lower part of the given range, highest spectral efficiency can be achieved with a moderate symbol rate.

The possibility to tune the FSR should also be considered. Of course, each comb source could be engineered to match a certain frequency grid. However, many techniques to generate frequency combs offer only a limited range for changing the FSR during operation. The tuning range might even be too small to compensate for fabrication tolerances. Therefore a tunable FSR, as offered by the modulator-based approaches, see Section 2.1.2, opens up many opportunities. First, a tunable FSR relaxes the tolerance on the pass-bands of the DEMUX filter. Second, a tunable FSR reduces development and production cost as only one comb source needs to be developed rather than having to adapt the comb source for each FSR individually. Third, a second comb could be tuned to match the first one for providing all the LO tones at the receiver.

Power and OCNR

The power and noise of the comb lines typically vary across the spectrum, see Fig. 2.3. This has some important consequences for using a comb source in data transmission. In the following paragraphs, we discuss the effects of having a low power per comb line and a non-flat spectral envelope. We describe the measurement of the OCNR, and we introduce a minimum OCNR that a comb line must have in order to be useful for data transmission. The OCNR metric does not consider phase noise, which is treated separately in Section 2.2.2.

A high Tx power is required to overcome the limiting receiver sensitivity. Hence, the lower power a comb line has, the more amplification it requires, thereby adding a larger amount of amplified spontaneous emission (ASE) and degrading the OCNR. Comb lines with too little power are useless for data transmission. The power per comb line is limited by the technology used for comb generation, and by the optical components following the comb source, e.g., the DEMUX filter.

The lower limit for the power per comb line is evaluated in each experiment individually. It depends on many system parameters as modulation format, power budget including margins, distance to and sensitivity of the receiver. This thesis searches for the highest possible data rate for any given comb source. To this end the largest possible amount of comb lines is used, and for each of them the modulation format with highest possible spectral efficiency is chosen. The transmitted distance in each case is at least 75 km.

For a comparable performance of all WDM data channels, an equalization (flattening) of the comb line powers is required, which is done by attenuating the stronger comb lines to the power level of the weakest one. Of course, the weakest one still needs to be strong enough to support the desired modulation format and reach. The loss induced by the flattening needs to be compensated by a subsequent amplification. Therefore, the flatness of the comb's spectral envelope is a very important metric. Many combs also feature significant power in their spectral wings, where the individual comb lines are too weak to be used for data transmission. These lines are suppressed by the equalizer to prevent saturation of the subsequent amplifiers. From an efficiency point of view this power is lost and degrades the wall-plug efficiency of the comb source. It would be therefore desirable that the comb's spectral envelope resembles a rectangle. In our experiments a programmable filter is used for equalization. The filter profile is adapted for each experiment individually as explained in the respective sections.

The OCNR relates the power P_m of the comb line with index m to the noise power $P_{N,m}$ at the frequency of the comb line where the measured noise power is normalized to an arbitrary reference bandwidth B_{ref}. A reasonably good estimate of the OCNR can be obtained from the measured optical spectrum. To this end, the carrier power P_m is measured in a bandwidth larger than the carrier linewidth, and the noise power $P_{N,m}$ at the carrier frequency is estimated by interpolating the measured noise power spectral density S_N in the middle between the carrier and the neighboring comb lines, see blue circles in Fig. 2.3. For comparing the measured OCNR values among different combs, the noise power, as measured in a bandwidth B_{RBW}, is normalized to B_{ref}. In this work, instead of defining B_{ref}, we use the optical carrier-to-noise-density power ratio (OCN_0R).

$$\text{OCN}_0\text{R}=\frac{P}{S_N} \tag{1.2}$$

An OCN_0R value of 125 dB Hz corresponds to an OCNR of approximately 24 dB when taking the noise power within a reference bandwidth of 0.1 nm at a center wavelength of about 1.5 µm as done, e. g., in [112], [113]. Note that some authors also use the term optical signal-to-noise power ratio (OSNR) synonymous to OCNR even if there is an unmodulated carrier only rather than a data signal.

Throughout this thesis we distinguish between the $OSNR_{ref}$, where the measured noise power is normalized to an arbitrary reference bandwidth B_{ref}, and the OSNR, which relates the signal power in the optical signal bandwidth B_O to the noise power in the same bandwidth. The relation between OSNR and $OSNR_{ref}$ is [7], [114], [115]

$$\mathrm{OSNR} = \frac{2B_{\mathrm{ref}}}{pB_O}\mathrm{OSNR}_{\mathrm{ref}}. \tag{1.3}$$

It is common to choose $\lambda_{\mathrm{ref}} = 0.1\,\mathrm{nm}$, which corresponds to $B_{\mathrm{ref}} = 12.5\,\mathrm{GHz}$ at a central wavelength of about 1.5 µm. For a single-polarization signal $p = 1$ holds, while $p = 2$ is valid for a polarization-multiplexed signal.

For finding a lower limit for the OCNR of the comb lines, the OSNR requirements for QPSK and 16QAM are discussed in the following. Fig. 2.4 shows BER versus OSNR for the two modulation formats, calculated according to [116]. Note that the formula given in [116] uses the electrical SNR per bit, which is converted to the electrical SNR assuming "uncoded data", i.e., without adding redundancy to the information bits [7]. We further assume that the electrical SNR is equal to the OSNR. This is true for optically amplified links if the dominating noise at the receiver is due to signal-spontaneous beat noise, and if the bandwidth of the optical noise filter is twice as large as the electrical receiver bandwidth [117].

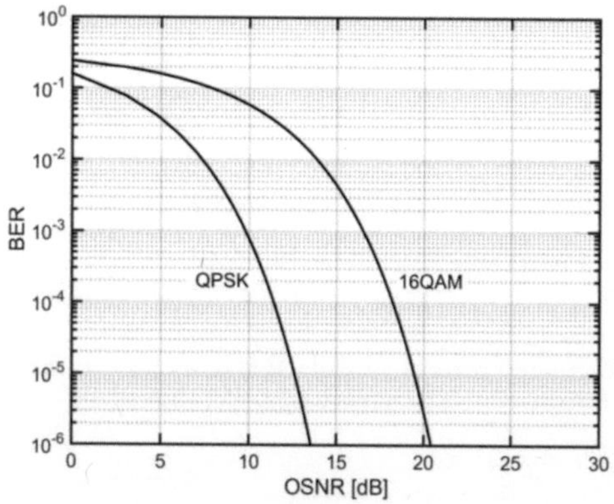

Fig. 2.4: Bit error ratio (BER) vs. optical signal-to-noise power ratio (OSNR) for quadrature phase shift keying (QPSK) and 16-state quadrature amplitude modulation (16QAM).

In order to reach a BER below 10^{-3}, a minimum OSNR of 9.8 dB and 16.5 dB is required for QPSK and 16QAM, respectively. For a polarization-multiplexed sinc-shaped signal with a symbol rate of $R_s = 18\,\text{GBd}$, i.e., $B_O = 18\,\text{GHz}$, this corresponds to a minimum OSNR_{ref} of 11.4 dB and 18.1 dB, respectively. A lower limit for the OCN_0R can be given by choosing $B_{\text{ref}} = 1\,\text{Hz}$. This leads to a minimum OCN_0R of 112.4 dB Hz required for QPSK, and 119.1 dB Hz needed for 16QAM in order to achieve a BER below 10^{-3}. According to the discussion in Appendix C.2 the OCN_0R-limits can be converted to limits for the relative intensity noise (RIN). Normalized to a bandwidth $B = 1\,\text{Hz}$, the limits are $-109.4\,\text{dB}\,\text{Hz}^{-1}$ and $-116.1\,\text{dB}\,\text{Hz}^{-1}$ for QPSK and 16QAM, respectively.

2.2.2 Linewidth

Phase noise determines the linewidth of an oscillator, as amplitude noise is reduced by the oscillators inherent amplitude control mechanism, see discussion in Appendix C.1. For coherent data transmission, the phase uncertainty of the carrier as well as of the local oscillator translate to an uncertainty of the azimuthal position of the measured symbols in the constellation diagram. Since the impact of phase noise is determined by the phase noise accumulated over the symbol duration, the linewidth requirements decrease with the symbol rate.

The tolerance towards phase noise depends on the modulation format and on the phase recovery algorithms that are implemented at the receiver digital signal processing [12]. A collection of literature values for the required product of linewidth and symbol duration is given in Table 2-1. These values are evaluated at an SNR penalty of 1 dB at a given pre-FEC BER of 10^{-3}. Multiplying the respective number with the targeted symbol rate gives the minimum required combined linewidth of carrier and LO. For example, with a symbol rate of 18 GBd, the linewidth requirements according to [118] are 4.7 MHz for QPSK and 1.1 MHz for 16QAM, respectively.

Table 2-1: Minimum required product of linewidth times symbol duration for QPSK and 16QAM for an SNR penalty of 1 dB at a given pre-FEC BER of 10^{-3}.

QPSK	16QAM	Reference
2.6×10^{-4}	6.0×10^{-5}	[118]
4.1×10^{-4}	1.4×10^{-4}	[119]
2.7×10^{-4}	2.9×10^{-5}	[120]

There are multiple options to measure the linewidth of comb lines. The easiest one would be to do a direct spectroscopy. However, even the highest resolution optical spectrum analyzers that are currently available can only resolve a linewidth of several tens of MHz. For a better resolution, the comb lines need to be down-converted and measured with an electrical spectrum analyzer. This can be done with a laser that has a significantly smaller linewidth than the comb line under investigation, see Section 4.2.3 or [121] for more detailed information about this measurement technique. The other option is to use the comb line itself for down-conversion. In this so-called self-heterodyne technique [122] the comb line is mixed with a frequency-shifted and de-correlated version of itself. The frequency shift allows using an electrical spectrum analyzer at an intermediate frequency larger than zero. The line as measured in the electrical domain is twice as broad as the original optical comb line.

3 Tbit/s data transmission with a gain-switched comb source

The contents of this chapter correspond to publication [J2], except for section 5 of [J2], which has been extended and is located in Chapter 1. A part of the results were presented at a scientific conference [C13].

Flexible terabit/s Nyquist-WDM super-channels using a gain-switched comb source

Optics Express; Volume 23; Issue 2; pp. 724-738; January 2015
DOI: 10.1364/OE.23.000724

J. Pfeifle[1], V. Vujicic[2], R. T. Watts[2], P. C. Schindler[1], C. Weimann[1], R. Zhou[2], W. Freude[1,3], L. P. Barry[2], C. Koos[1,3]

[1]*Institute of Photonics and Quantum Electronics (IPQ), Karlsruhe Institute of Technology (KIT), Germany*

[2]*The Rince Institute, School of Electronic Engineering, Dublin City University (DCU), Ireland*

[3]*Institute of Microstructure Technology (IMT), Karlsruhe Institute of Technology (KIT), Germany*

Tbit/s super-channels are likely to become the standard for next-generation optical networks and optical interconnects. A particularly promising approach exploits optical frequency combs for super-channel generation. We show that injection locking of a gain-switched laser diode can be used to generate frequency combs that are particularly well suited for Tbit/s super-channel transmission. This approach stands out due to its extraordinary stability and flexibility in tuning both center wavelength and line spacing. We perform a series of transmission experiments using different comb line spacings and modulation formats. Using 9 comb lines and 16QAM signaling, an aggregate line rate (net data rate) of 1.296 Tbit/s (1.109 Tbit/s) is achieved for transmission over 150 km of standard single mode fiber (SSMF) using a bandwidth of 166.5 GHz, which corresponds to a (net) spectral efficiency of 7.8 bit/s/Hz (6.7 bit/s/Hz). The line rate (net data rate) can be boosted to 2.112 Tbit/s (1.867 Tbit/s) for transmission over 300 km of SSMF by using a bandwidth of 300 GHz and QPSK modulation on the weaker carriers. For the reported net data rates and spectral efficiencies, we assume a variable overhead of either 7 % or 20 % for forward-error correction depending on the individual sub-channel quality after fiber transmission.

3.1 Introduction

Interfaces operating at 400 Gbit/s or 1 Tbit/s are foreseen as the next standards after 100 Gbit/s Ethernet [11], [35], [123]. In this context, optical super-channels are promising candidates, combining a multitude of sub-channels in a wavelength-division multiplexing (WDM) scheme, while each sub-channel operates at a moderate symbol rate that complies with currently available CMOS driver circuitry. Typically the super-channels use spectrally efficient advanced modulation formats such as quadrature phase shift keying (QPSK) or 16-state quadrature amplitude modulation (QAM) in combination with advanced multiplexing schemes such as orthogonal frequency-division multiplexing (OFDM) [50], [51], [124] or Nyquist-WDM [44], [52], [125], [126]. The performance of these transmission schemes depends heavily on the properties of the optical source, in particular on the number of lines, the power per line, the optical carrier-to-noise power ratio (OCNR), the optical linewidth, and on the relative intensity noise (RIN). In addition, tunability of emission wavelength and line spacing are important features providing flexibility of the transmission scheme. Tbit/s super-channel transmission has been demonstrated before, using different approaches for realizing the optical source. These approaches include ensembles of independent lasers [127], [128], single-laser concepts with sidebands generated by external modulation [J5], [129] or by Kerr comb generation in optically nonlinear high-Q microresonators [J4], and fiber-based spectral broadening [50], [52], [53], [113], [130]. Arrays of independent lasers offer flexibility and do not require filters for spectrally separating the comb lines. However, the achievable spectral efficiency (SE) is limited by the drift of the individual emission wavelengths and the associated guard bands. For modulator-based comb sources, the number of comb lines is generally limited by the achievable modulation depth [J5] unless complex arrangements of cascaded modulators with synchronized driving signals are used [129], [131]. Moreover, if large numbers of lines are to be generated, the power of the comb lines is limited by the power handling capacity and by the insertion loss of the modulator. Kerr comb sources can provide a large numbers of lines, but require sophisticated pumping schemes and the line spacing is inherently tied to the free spectral range of the underlying resonator. Broadband frequency combs for Tbit/s super-channels can also be generated by solid-state mode-locked lasers in combination with highly nonlinear fibers [50], [52], [112], or multi-stage parametric mixers [53], [113], [130]. However, both approaches require strong optical pumps and large interaction lengths in delicately arranged sequences of specialized optical fibers. Integration density is hence limited. While this might be acceptable when considering the ultra-high data rates achieved with these sources — several tens of Tbit/s — the associated higher power consumption is not very well suited for point-to-point links that require only a limited number of comb lines (order of 10) to form a Tbit/s super-channel.

In this work we demonstrate that gain-switched comb sources (GSCS) [58] can be used as an alternative approach to generate Tbit/s super-channels. These devices exploit injection locking of a gain-switched laser diode and feature both an electrically tunable free spectral range [100] and an electrically tunable center wavelength [59], good spectral flatness, high OCNR, low RIN and low optical linewidth. In Refs [132], [133] such a comb source has been used to transmit several channels operating at 10 Gbit/s and 40 Gbit/s, respectively, in a passive optical network (PON) with coherent detection. The spectral efficiency in these experiments amounted to 1 bit/s/Hz and 2 bit/s/Hz, respectively. Here we demonstrate six distinct Tbit/s super-channel experiments using polarization division multiplexed (PDM) QPSK and 16QAM in combination with Nyquist pulse shaping for efficient use of the available bandwidth. Transmission over various distances of standard single-mode fiber (SSMF) has been tested. We achieve an aggregate line rate (net data rate) of 1.296 Tbit/s (1.109 Tbit/s) after transmission over 150 km of SSMF in a bandwidth of 166.5 GHz [C13]. This corresponds to a (net) spectral efficiency of 7.8 bit/s/Hz (6.7 bit/s/Hz). When combining 16QAM on strong carriers and QPSK modulation on the weaker carriers, the line rate (net data rate) can be boosted to 2.112 Tbit/s (1.867 Tbit/s) and transmitted over 300 km using a bandwidth of 300 GHz.

3.2 The gain-switched comb source

Fig. 3.1(a) shows the experimental setup of the GSCS. Gain switching is achieved by driving a distributed feedback (DFB) slave laser diode with a large sinusoidal signal (24 dBm) at a frequency that corresponds to the desired line spacing, in combination with a DC bias current of approximately four times the threshold current ($I_{\mathrm{th}} \approx 12.5\,\mathrm{mA}$). Additionally, a master laser (Agilent N7711A) injects continuous-wave light into the slave laser via a polarization controller (PC) and an optical circulator. The master laser establishes coherence between subsequent pulses and transfers its characteristically low optical linewidth (80 kHz) to the individual modes of the comb [58], [59], and reduces the RIN of the comb modes [60]. The comb's center wavelength can be tuned by simultaneously adjusting the emission wavelength of the master laser and the temperature of the DFB slave. The polarization controller is used to align the polarization state of the injected light with that of the slave laser mode. By carefully selecting the wavelength detuning and the injected power, long-term (during several days) operational stability of the GSCS can be achieved without automated feedback control or manual adjustment of detuning and polarization. In our experiments, the injection power is set to approximately 7 dBm, and the slave laser is temperature-controlled at 25 °C.

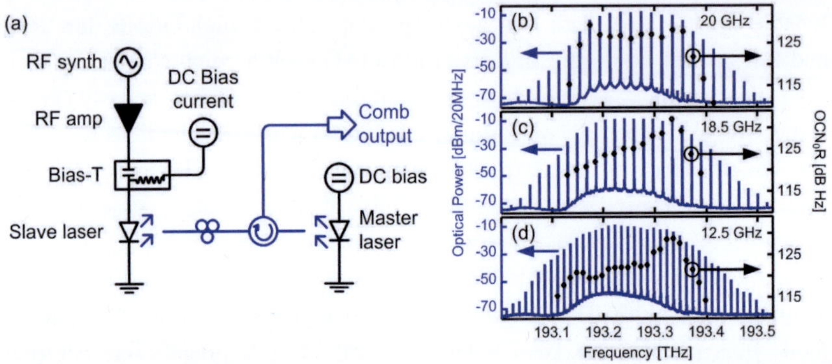

Fig. 3.1: Setup and spectra of GSCS. (a) Setup schematic of the GSCS. A frequency comb is generated by gain switching of the DFB slave laser. The line spacing is determined by the frequency of the RF drive signal. By injecting light from the master laser to the slave laser via a circulator, the low linewidth and RIN of the master laser are transferred to the comb lines. (b-d) Comb spectra (blue, left axis) measured at position "Comb output" with a line spacing of (b) 20 GHz, (c) 18.5 GHz, and (d) 12.5 GHz (RBW 20 MHz) and optical carrier-to-noise-density power ratio (OCN_0R – black, right axis) for the comb lines used for the data transmission experiments.

We investigate the GSCS with three different line spacings, 20 GHz, 18.5 GHz and 12.5 GHz, by appropriately changing the frequency of the RF synthesizer. The respective GSCS spectra are depicted in Fig. 3.1(b)–(d) on the left-hand axis (blue). The power levels correspond to the actual power of the comb measured at position "Comb output" in Fig. 3.1(a). The total power in all three cases amounts to +3.5 dBm. On the right-hand axis of Fig. 3.1(b)–(d) we depict the optical carrier-to-noise-density power ratio (OCN_0R) of the comb lines that are used for the data transmission experiments. The OCN_0R is defined by the ratio of the power of the unmodulated carrier to the underlying noise power in a bandwidth of 1 Hz. Note that some authors use a reference bandwidth of 0.1 nm for evaluating the noise power. An OCN_0R value of 125 dB Hz corresponds to an OCNR (by some authors named OSNR) of approximately 24 dB, when measuring the noise in a reference bandwidth of 0.1 nm as done, e. g., in [112], [113]. The origin of the noise floor in the GSCS spectra is discussed in [134], where the trade-off between flatness and high-frequency FM noise was elaborated. Here, we concentrate on "flat combs" [134], accepting the noise penalty in favor of a higher number of sub-channels and hence higher aggregate data rates.

Note that the optical bandwidth of all three combs in Fig. 3.1(b)–(d) is the same. The bandwidth is dictated by the intrinsic dynamics of photons and electrons within the slave laser. Choosing a smaller line spacing will yield a larger number of lines. With respect to the possible number and the power of lines, the GSCS competes with frequency combs generated with a single electro-optic modulator. The main advantage of the GSCS is its inherent stability without any bias control. Even the polarization of the master laser is uncritical once it is set, because a slight polarization change results in a minor reduction of the locking range only. The temperature control of the master laser would be required anyway for the laser in a modulator-based comb generator. Note that the current setup for comb generation comprises discrete components. Further cost and complexity reduction as well as a stability improvement can be obtained by a monolithic integration of master and slave lasers [135]. For an integrated GSCS, the total power consumption can be estimated to be approximately 10 W, see Table 3-1. The total optical output power then amounts to typically 3.5 dBm for the entire comb. For some applications, the output power needs to be boosted to values of 10 – 17 dBm per line. This can be achieved by appropriate amplifier stages adding at most 10 W to the total power consumption.

Table 3-1: Power consumption estimation for a GSCS using integrated components

Component	Estimated power consumption
Two-channel laser driver	0.7 W[a]
Temperature controllers	2×1.75 W[b]
RF source	1.3 W[c]
RF amplifier	4.5 W[d]

[a]http://www.teamwavelength.com/products/product.php?part=147&view=specs#tabs, assuming both lasers are operated at 80 mA.
[b]http://www.maximintegrated.com/en/products/power/switching-regulators/MAX8520.html
[c]https://www.hittite.com/products/view.html/view/HMC807LP6CE
[d]http://www.shf.de/wp-content/uploads/datasheets/datasheet_shf_100bp.pdf

3.3 Super-channel generation and characterization

For a given line spacing, super-channel capacity is dictated by two parameters: First, the number of carriers that can be derived from the comb source, which defines the number of sub-channels, and second, the power levels and OCNR of the respective carriers that determine the modulation formats to be used on each sub-channel provided that the carrier linewidth is sufficiently low. For a more detailed discussion, see Section 2.2. As a matter of fact, the comb line power is highest in the center of the GSCS spectrum and decreases towards the periphery. In practical transmission systems, however, equal power distribution among all involved sub-channels is desired. This requires attenuation of the center comb lines relative to the power of the outer ones, and subsequent amplification to overcome insertion and modulation losses of the transmitter. The associated amplified spontaneous emission (ASE) noise limits the performance of the entire super-channel and is hence a crucial parameter when designing comb-based transmission systems. In the following experiments, we seek to maximize super-channel performance for a given transmission system, and to investigate the trade-off between spectral efficiency and transmission reach for the case of the GSCS.

The experimental setup to emulate a super-channel transmitter (Tx) and receiver (Rx) is depicted in Fig. 3.2. We use a programmable filter (Finisar WaveShaper) to equalize the power in the comb lines and to reject outer comb lines that feature too little power for sufficient modulation. The comb lines are then dis-interleaved into two sets of sub-carriers (odd and even). For the 20 GHz and 18.5 GHz line spacing, the dis-interleaving can be directly performed by the programmable filter, while for the 12.5 GHz comb, a commercially available interleaver (Optoplex Corp.) is used. The two sets of sub-carriers are amplified by two nominally identical EDFA operated in constant output power mode set to 16 dBm. Note that the equalization of the comb results in a reduction of the input power to these amplifiers. This limits the flattening of the comb lines as stronger equalization requires higher amplification afterwards, and hence more ASE noise adds to the comb lines. While this has a negative effect on the OCNR of the carriers, it helps to ensure comparable input power levels for the transmission EDFA and at the receiver input, so that the various transmission experiments can be compared.

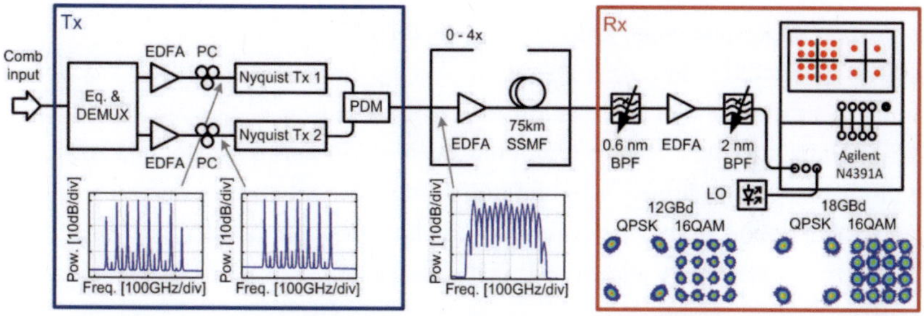

Fig. 3.2: Schematic of the GSCS Tbit/s super-channel transmitter and coherent receiver. The frequency comb is equalized and de-multiplexed into odd and even comb lines. The two sets of carriers are amplified, modulated with independent, sinc-shaped Nyquist signals and then combined in a polarization division multiplexing (PDM) scheme. Insets show the optical spectra measured using a 0.01 nm resolution bandwidth. The super-channel is either sent directly to the receiver or transmitted through up to four spans of 75 km standard single mode fiber (SSMF) with an EDFA before each span. At the receiver, the desired sub-channel is selected by a band-pass filter (BPF), amplified and coherently detected using a narrow-linewidth laser as a local oscillator (LO). The signal is recorded and analyzed using an optical modulation analyzer (Agilent N4391A). The presented constellation diagrams are obtained using an ECL as a carrier, and serve as reference measurements. The phase error that can be particularly seen in the 12 GBd constellation diagrams as well as the stronger noise cloud for the 18 GBd measurements is attributed to our transmitter hardware.

The two sub-carrier sets are then independently modulated using an IQ modulator driven by band-limited Nyquist pulses that are generated by a proprietary multi-format transmitter (Nyquist-Tx) [44], [136]. Polarization division multiplexing (PDM) is emulated by splitting the combined outputs of the two Nyquist-Tx into two paths, which are recombined after different delays to form two orthogonal polarization states in a SSMF. The insets show exemplary optical spectra, RBW 0.01 nm, of the separate sub-carrier sets and the combined Tbit/s data stream for the case of the 20 GHz comb and 18 GBd QPSK modulation. When using the waveshaper for dis-interleaving and flattening, we find that the crosstalk, i. e., the ratio of the power of a comb line at its allocated waveshaper output port and the residual power of the same comb line at the other output port is actually a function of the attenuation supplied to this particular comb line. We find crosstalk attenuation levels of (19.9±4.4) dB for the QPSK and (29.1±6.4) dB for the 16QAM experiments. We do not observe a correlation between crosstalk attenuation and the performance of the sub-channels, and hence conclude that our results are not limited by crosstalk effects, This finding is in accordance to the one reported in [137], where a SNR penalty of less than 1 dB is predicted for these levels of crosstalk.

The super-channels were either sent directly to the Rx for back-to-back (B2B) characterization, or transmitted over up to four spans of 75 km SSMF with an erbium-doped fiber amplifier (EDFA) before each span. The launch power into the transmission fibers has been set to approximately –3 dBm per sub-channel, providing a good compromise between nonlinear signal impairments and noise. As a local oscillator (LO) we use an external cavity laser (ECL) specified with a linewidth less than 100 kHz. The signals were analyzed using an optical modulation analyzer (Agilent N4391A) and with offline processing. We perform digital brick-wall filtering, polarization demultiplexing, dispersion compensation, and adaptive equalization before evaluating the bit-error-ratio (BER) and the error vector magnitude (EVM_m) of each sub-channel individually. The EVM subscript "m" indicates the normalization of the EVM to the maximum power of the longest ideal constellation vector.

As a performance measure of the system we test both modulation paths individually with a narrow-linewidth ECL that is comparable to the ECL used as local oscillator. With 18 GBd modulation we achieve an average EVM_m of 10.5 % and 9.8 % for QPSK and 16QAM, respectively, while an average EVM_m of 8.5 % and 7.5 % is obtained for 12 GBd modulation, respectively. The dependence of the performance on symbol rate is attributed to the limited bandwidth of our anti-aliasing filters after digital-to-analog conversion.

Assuming that the signal is impaired by additive white Gaussian noise only, a direct relation from the EVM_m to the BER can be established [114], [115]. In our experiments this assumption is not perfectly fulfilled as the constellations show a small signature of phase noise. As we see this behavior also for the reference measurements with an ECL, see insets in Fig. 3.2, we attribute this to our transmitter hardware rather than to the linewidth of the optical source. Nevertheless we present our results using the EVM_m metric, since for the QPSK experiments we generally did not measure a high-enough number of errors within the length of one recording. For the 16QAM experiments we give both the EVM_m and the BER. One can see that the EVM_m slightly overestimates the signal quality, which we attribute to the small phase error introduced by our transmitter hardware as well as the fact that our EVM_m measurement is non-data-aided while the relation in [114] assumes a data-aided measurement of the EVM_m.

Another interesting metric would be the sub-channel OSNR. However, such an evaluation is not easily possible in our de-multiplexer setup: Due to narrowband filtering of the individual comb lines using a passband width on the order of 10 GHz, the filtered ASE background is strongly non-uniform, and we can hence not any more estimate the noise power density at the carrier wavelength from a measurement taken in the center between two carriers. We may, however, estimate the sub-channel OSNR from the OCN_0R measured at the output of the GSCS. For the 20 GHz comb, the OCN_0R is relatively uniform and amounts to approximately 127 dB Hz, which would translate into an OSNR of roughly 26 dB in a reference bandwidth of 0.1 nm after modulation and back-to-back reception.

3.4 Experimental results

3.4.1 Experimental parameters and comparison of super-channels

We investigate six different super-channel architectures that are based on three different line spacings, 20 GHz, 18.5 GHz and 12.5 GHz, and two different modulation formats, QPSK and 16QAM. The results of all six experiments are summarized in Table 3-2. For each super-channel we take as many comb lines as possible, each carrying a Nyquist-WDM sub-channel. Comb lines that were too weak for the respective modulation format were suppressed by the equalization filter (equalizer "Eq." in Fig. 3.2) prior to recording the data presented here. For the super-channels that were derived from the 20 GHz and the 18.5 GHz comb we use a symbol rate of 18 GBd and hence obtain a sub-channel line rate of 72 Gbit/s for PDM-QPSK modulation and 144 Gbit/s for PDM-16QAM. A symbol rate of 12 GBd was used for the super-channels based on the 12.5 GHz comb leading to a sub-channel line rate of 48 Gbit/s and 96 Gbit/s for the two modulation formats employed. For all symbol rates, the clock rate of the digital-to-analog converters (DAC) of the Nyquist-Tx was kept constant at 24 GHz, while the oversampling factor q for generating the sinc-shaped output pulses was adapted to $q = 4/3$ for 18 GBd and $q = 2$ for 12 GBd.

Table 3-2: Summary of all super-channels

Line spacing [GHz]	Mod. format	# of sub-ch.	Sub-ch. line rate [Gbit/s]	Aggr. line rate [Tbit/s]	Occ. bandw. [GHz]	Net aggr. data rate [Tbit/s]	Net SE [bit/s/Hz] (tr. dist.)
20	PDM-QPSK	13	72	0.936	260	0.875	3.4 (300 km)
	PDM-16QAM	8	144	1.152	160	1.047	6.5 (150 km)
18.5	PDM-QPSK	15	72	1.080	277.5	1.009	3.6 (300 km)
	PDM-16QAM	9	144	1.296	166.5	1.109	6.7 (150 km)
12.5	PDM-QPSK	24	48	1.152	300	1.077	3.6 (300 km)
	PDM-QPSK / PDM-16QAM	4 / 20	48 / 96	2.112	300	1.867	6.2 (300 km)

In Table 3-2, the aggregate line rate is obtained by multiplication of the number of sub-channels with the sub-channel line rate, and the bandwidth of the super-channel is given by the product of sub-channel number with the comb line spacing. For calculating the net aggregate data rate one needs to take into account the overhead for forward error correction (FEC). FEC schemes with 7 % overhead can cope with BER of up to 4.5×10^{-3} [138], a requirement which was fulfilled for all B2B experiments. For some transmission experiments, this BER threshold is exceeded as discussed in the subsequent sections. In these cases, some of the sub-channels require more advanced FEC codes with a larger overhead of, e.g. 20 %, having a higher BER limit of, e.g., 1.5×10^{-2} [139]. We calculate the net aggregate data rate for each super-channel after the longest tested transmission distance by taking into account the respective overhead for each sub-channel individually. The net SE is then calculated from the ratio of the net aggregate data rate and the occupied bandwidth. In the parenthesis we indicate the actual fiber transmission distance over which this spectral efficiency has been achieved.

3.4.2 Tbit/s super-channels with an 18.5 GHz GSCS

For the highest spectral efficiency we operate the Nyquist-Tx at a symbol rate of 18 GBd and choose a comb line spacing of 18.5 GHz, see spectrum of Fig. 3.1(c). We select a total of 15 lines using the equalization filter and flatten the spectrum to the level of the weakest carrier. Fig. 3.3(a) shows the spectrum of the super-channel using PDM-QPSK modulation with a total aggregate line rate of 1.08 Tbit/s, measured in a resolution bandwidth of 0.01 nm. For each sub-channel the measured EVM_m, averaged over both polarizations, is presented in Fig. 3.3(b) for the B2B case and for transmission distances of 75 km, 150 km, 225 km, and 300 km. The transmission lengths are distinguished by different symbols, colors, and by an offset in the horizontal direction. Note that we could not find enough errors in the recorded number of 4,500,000 bits for evaluating the BER reliably. Therefore we conclude that the BER of all sub-channels are clearly below the threshold for second-generation forward-error correction (FEC) with 7 % overhead, given by a BER of 4.5×10^{-3}, which corresponds to an EVM_m of 38.3 % [114]. We hence obtain a net super-channel capacity of 1.009 Tbit/s in a bandwidth of 278 GHz, corresponding to a net spectral efficiency of 3.6 bit/s/Hz. Selected constellations diagrams of the B2B-experiments are displayed in Fig. 3.3(c). We compare two sub-channels, one at the border and one at the center of the super-channel spectrum. They perform similar for both polarizations.

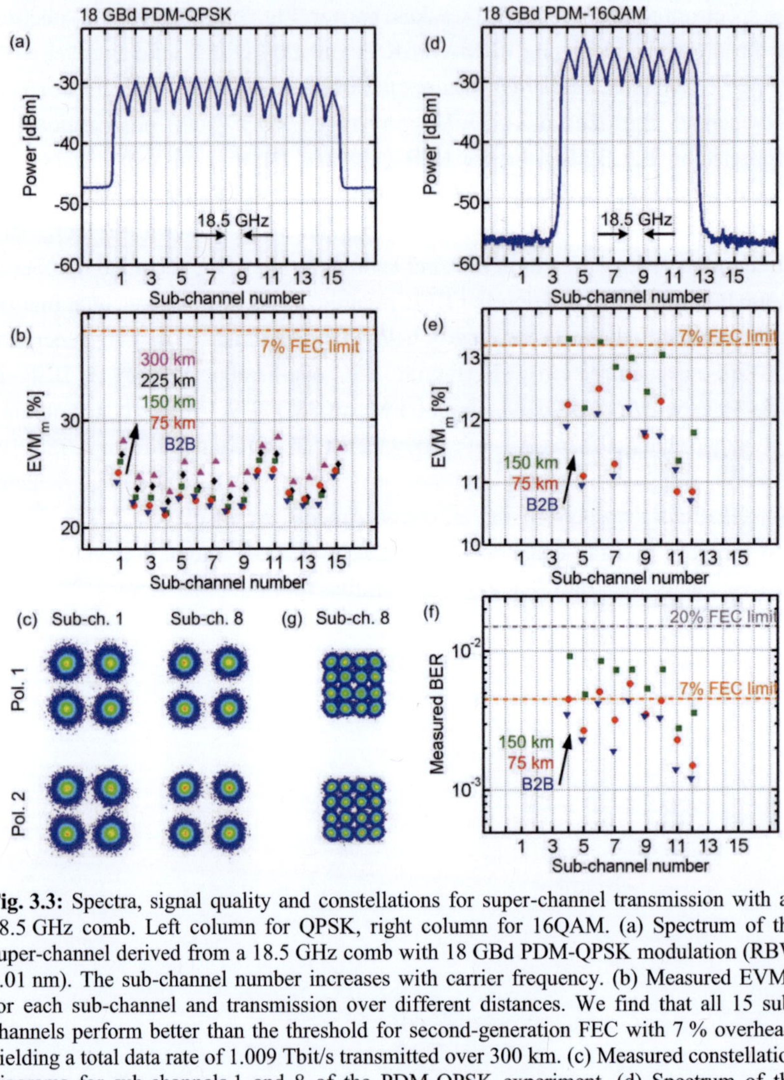

Fig. 3.3: Spectra, signal quality and constellations for super-channel transmission with an 18.5 GHz comb. Left column for QPSK, right column for 16QAM. (a) Spectrum of the super-channel derived from a 18.5 GHz comb with 18 GBd PDM-QPSK modulation (RBW 0.01 nm). The sub-channel number increases with carrier frequency. (b) Measured EVM_m for each sub-channel and transmission over different distances. We find that all 15 sub-channels perform better than the threshold for second-generation FEC with 7 % overhead, yielding a total data rate of 1.009 Tbit/s transmitted over 300 km. (c) Measured constellation diagrams for sub-channels 1 and 8 of the PDM-QPSK experiment. (d) Spectrum of the super-channel derived from a 18.5 GHz comb with 18 GBd PDM-16QAM (RBW 0.01 nm). (e-f) Measured EVM_m and BER for each sub-channel and transmission over different distances. We find that 2 (7) sub-channels perform better than the FEC threshold with 7 % (20 %) overhead, yielding a aggregate net data rate of 1.077 Tbit/s transmitted over 150 km. (g) Measured constellation diagrams for sub-channel 8 of the PDM-16QAM experiment.

The bandwidth requirement can be reduced by a factor of two when upgrading the system from QPSK to 16QAM signaling. This doubles the line rate per channel but requires a higher optical signal-to-noise power ratio (OSNR). To this end we increase the equalization power level for flattening the comb, i. e., the average attenuation is decreased, which leads to carriers with a higher power. However, only fewer comb lines contribute the required power, hence the number of sub-channels is reduced from 15 to 9. Because a higher power level is input to the EDFA, the amplification and the ASE contribution is lower, leading to a larger OCNR. Additionally, the smaller attenuation results in a reduced crosstalk of the waveshaper, see Section 3.3, third paragraph. Spectrum, EVM_m and measured BER are presented in Fig. 3.3(d)–(f). For short transmission distances (B2B), the BER stays below the standard FEC limit for 7 % overhead. For a transmission distance of 150 km, seven of the nine sub-channels exceed the BER mark for standard FEC, and implementations with larger overhead, i. e., 20 % overhead, become necessary. The remaining two sub-channels fall below the FEC threshold with 7 % overhead even after transmission over 150 km. The aggregate line rate (net data rate) of this super-channel amounts to 1.296 Tbit/s (1.109 Tbit/s), corresponding to a (net) spectral efficiency of 7.8 bit/s/Hz (6.7 bit/s/Hz). This is among the highest values achieved for 16QAM in Tbit/s super-channels.

In Fig. 3.3(g) we show the constellation diagram of the central sub-channel 8 in two polarizations. Note that the smaller noise clouds as compared to Fig. 3.3(c) (QPSK) for the same sub-channel is a direct consequence of the higher power level of the flattened comb lines in the 16QAM case, which yields a higher OSNR of all sub-channels.

3.4.3 Tbit/s super-channels with a 20 GHz GSCS

A more robust super-channel can be generated by increasing the guard band. To this end, we increase the line spacing of the comb to 20 GHz, see spectrum in Fig. 3.1(b), while keeping the same symbol rate at the Nyquist-Tx. We select again 15 comb lines, 13 of which we flatten to the same power level as in the 18.5 GHz experiment for better comparison. The outermost carriers do not reach this power level and hence their performance drops below that of the inner 13 carriers. This can be seen in the spectrum as well as in the EVM_m results for QPSK modulation, Fig. 3.4(a) and (b). Taking into account the inner 13 sub-channels, the super-channel aggregate line rate (net data rate) amounts to 0.936 Tbit/s (0.875 Tbit/s), with a (net) spectral efficiency of 3.6 bit/s/Hz (3.4 bit/s/Hz).

Selected constellation diagrams of the B2B-experiments are displayed in Fig. 3.4(c). We compare two sub-channels, one at the border of the comb (sub-channel 1, also representative for sub-channel 15), and the central sub-channel 8 (representative for sub-channels 2 to 14). Sub-channels 1 and 15 suffer from low carrier power and therefore from a small OSNR. However, the central sub-channels 2 to 14 have an EVM_m which is slightly better than that of the corresponding sub-channels of the 18.5 GHz experiment. We attribute this to the larger guard band that reduces inter-channel interference originating from imperfect Nyquist pulse shaping.

This fact becomes even more evident when comparing the 16QAM 20 GHz comb experiment Fig. 3.4(d)–(f) with the 18.5 GHz comb experiment in Fig. 3.3(d)–(f). We achieve 6 (2) sub-channels that are better than the 7 % (20 %) FEC limit after 150 km of fiber transmission. The aggregate line rate (net data rate) amounts to 1.152 Tbit/s (1.047 Tbit/s). The (net) spectral efficiency amounts to 7.2 bit/s/Hz (6.5 bit/s/Hz). These results must be compared to the case of the 18.5 GHz comb and illustrate the trade-off between super-channel capacity and reach: For a comparable capacity, we used 9 sub-channels, but only 2 (7) were better than the 7 % (20 %) FEC limit after transmission over 150 km. We conclude that transmission with the 20 GHz comb has the larger capacity for a reach, where the channels with a FEC overhead of 20 % would drop out.

In Fig. 3.4(g) we show the constellation diagram of the central sub-channel 8 in two polarizations. The smaller noise clouds as compared to Fig. 3.4(c) (QPSK) for the same sub-channel follow from the higher power level of the flattened comb, see Fig. 3.3(c) and (g).

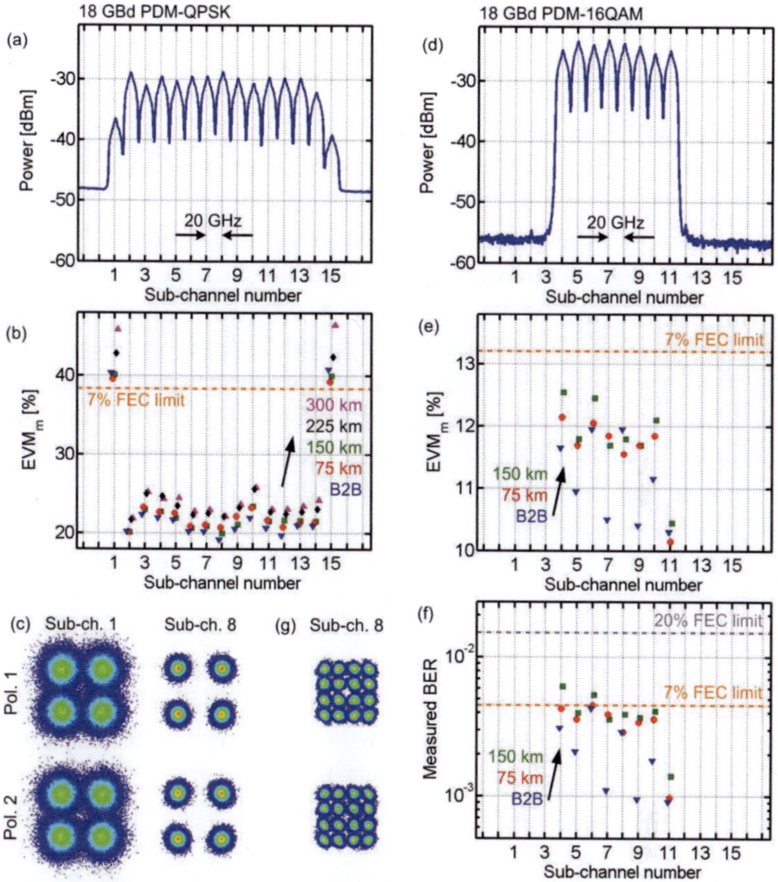

Fig. 3.4: Spectra, signal quality and constellations for super-channel transmission with a 20 GHz comb. Left column for QPSK, right column for 16QAM. (a) Spectrum of the super-channel derived from a 20 GHz comb with 18 GBd PDM-QPSK modulation (RBW 0.01 nm). The sub-channel number increases with carrier frequency. (b) Measured EVM$_m$ for each sub-channel and transmission over different distances. We find that 13 sub-channels perform better than the threshold for second-generation FEC with 7 % overhead, yielding a total data rate of 0.875 Tbit/s transmitted over 300 km. (c) Measured constellation diagrams for sub-channels 1 and 8 of the PDM-QPSK experiment. (d) Spectrum of the super-channel derived from a 18.5 GHz comb with 18 GBd PDM-16QAM (RBW 0.01 nm). (e-f) Measured EVM$_m$ and BER for each sub-channel and transmission over different distances. We find that 6 (2) sub-channels perform better than the FEC threshold with 7 % (20 %) overhead, yielding a total data rate of 1.047 Tbit/s transmitted over 150 km. (g) Measured constellation diagrams for sub-channel 8 of the PDM-16QAM experiment.

3.4.4 Tbit/s super-channels with a 12.5 GHz GSCS

In a last set of experiments, we take advantage of the tunability of the GSCS by adapting the line spacing and center frequency to the 12.5 GHz ITU-grid, see spectrum in Fig. 3.1(d). This enables the use of a commercial fixed interleaver to separate odd and even sub-carriers. The interleaver exhibits an excellent extinction of the odd carriers in the path of the even carriers, and vice versa. The programmable filter is then used for equalization only.

For the QPSK experiments, we use again the same equalization level as for the 18.5 GHz and the 20 GHz comb while reducing the symbol rate to 12 GBd. This leads to 24 sub-channels with EVM_m values well below the 7% FEC limit, see Fig. 3.5(a) for the super-channel spectrum and Fig. 3.5(b) for the EVM_m results. The four outermost channels do not reach the power level which was used for flattening, and hence drop in performance. The (net) capacity of this super-channel is 1.152 Tbit/s (1.077 Tbit/s), and the (net) spectral efficiency amounts to 3.8 bit/s/Hz (3.6 bit/s/Hz). Fig. 3.5(c) shows selected constellation diagrams for the QPSK experiment in the B2B case. The constellation diagrams represent peripheral (example: sub-channel 1) and central sub-channels (example: sub-channel 12), respectively.

We finally boost the super-channel aggregate line rate to 2.112 Tbit/s by increasing the equalization level and by admitting the weak outer comb lines, too. This enables 16QAM modulation on 20 sub-channels in the center, while 4 of the outer sub-channels still enable QPSK operation below the 7% FEC threshold, see Fig. 3.5(d)–(f). The aggregate net data rate amounts to 1.867 Tbit/s with a net spectral efficiency of 6.2 bit/s/Hz for transmission over 300 km. Fig. 3.5(g) shows the PDM-16QAM constellation diagram of a central 16QAM sub-channel.

Comparing the performance of the individual sub-channels for both super-channels with the 12.5 GHz GSCS it is possible to observe a slight reduction of the EVM_m and the BER with increasing sub-channel number. This performance dependence on the sub-channel number seems to be correlated with the variations of the OCN_0R of the comb lines as shown in Fig. 3.1(d). Note that, also the 18.5 GHz GSCS exhibits strong variations of the OCN_0R, but they do not seem to be correlated with the respective EVM_m. We believe that in contrast to operation at a symbol rate of 12 GBd, the cut-off frequency of the anti-aliasing filters becomes effective at the present symbol rate of 18 GBd (as explained in Section 3.3 Paragraph 5), and this dominates the OCN_0R-related performance degradation.

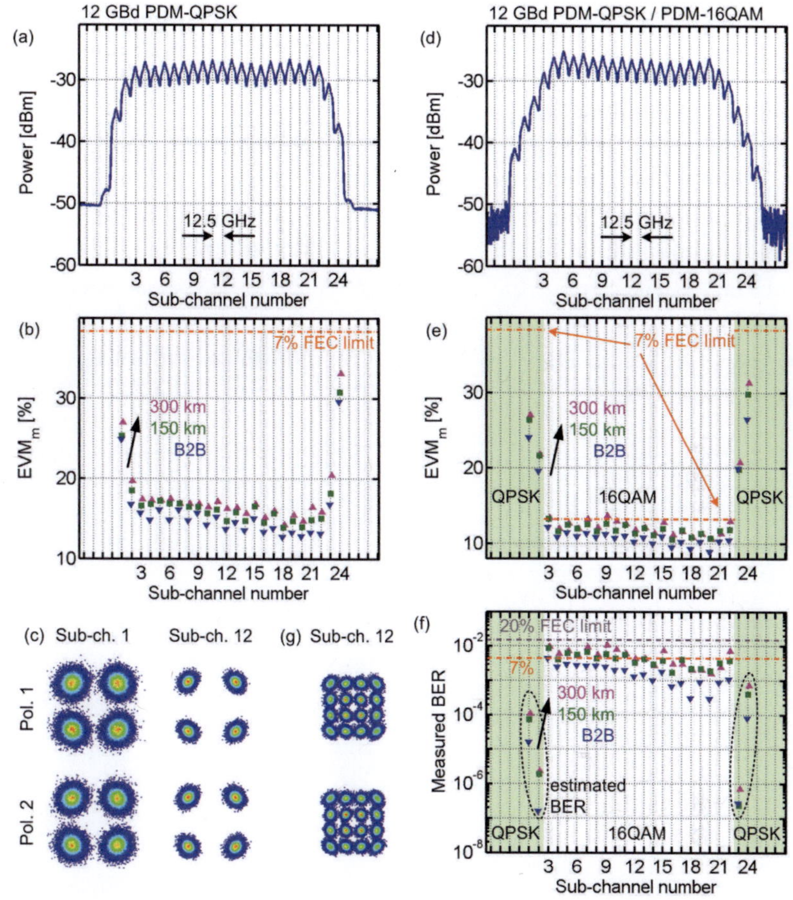

Fig. 3.5: Spectra, signal quality and constellations for super-channel transmission with a 12.5 GHz comb. Left column for QPSK, right column for mixed QPSK / 16QAM. (a) Spectrum of the super-channel derived from a 12.5 GHz comb with 12 GBd PDM-QPSK modulation (RBW 0.01 nm). The sub-channel number increases with carrier frequency. (b) Measured EVM$_m$ for each sub-channel and transmission over different distances. We find that all 24 sub-channels perform better than the threshold for second-generation FEC, yielding a total data rate of 1.077 Tbit/s transmitted over 300 km. (c) Constellation diagrams for the sub-channels 1 and 12. (d) Spectrum of the super-channel derived from a 12.5 GHz comb with 12 GBd PDM-16QAM (RBW 0.01 nm). (e-f) Measured EVM$_m$ and BER for 12 GBd PDM-16QAM on the inner comb lines (white background) and PDM-QPSK on the outer lines (green background), again for different propagation distances. We find that 13 (11) sub-channels (9 (11) channels using PDM-16QAM and 4 channels using PDM-QPSK) perform better than the threshold for FEC with 7 % (20 %) overhead, yielding a total data rate of 1.867 Tbit/s, transmitted over 300 km. (g) Measured constellation diagrams for sub-channel 12 of the PDM-QPSK / PDM-16QAM experiment.

3.5 Summary and conclusion

In summary, we show that GSCS are well suited for Tbit/s super-channel generation. We investigate six different super-channel architectures with different carrier spacings and modulation formats, and evaluate their performance for transmission over different distances. The highest (net) spectral efficiency, 7.8 bit/s/Hz (6.7 bit/s/Hz), is obtained for a GSCS with a carrier spacing of 18.5 GHz and 18 GBd PDM-16QAM, transmitted over 150 km. The data capacity transmitted over 300 km of SSMF can be increased to 1.867 Tbit/s by reducing the line spacing along with the symbol rate and by adapting the modulation format to the OSNR performance of the respective sub-channel.

4 Tbit/s data transmission with a quantum-dash mode-locked laser diode

Quantum-dash mode-locked laser diodes (QD-MLLD) are interesting comb sources because they are extremely compact and energy-efficient. They also feature more comb lines than the GSCS discussed in the previous chapter. In the framework of the European research program BIG PIPES[9] this comb source is investigated for data center applications. The laser chips are designed and fabricated by III-V Lab and Laboratoire de Photonique et Nanostructures, CNRS UPR20 in Marcoussis, France. In this chapter, the progress towards using this comb source for coherent data transmission is discussed. Section 4.1 introduces the device structure and the optical spectrum. Section 4.2 describes the experimental characterizations of the laser in view of its usefulness for coherent data transmission. These characterizations reveal that the optical linewidth of the comb lines is too large, making additional efforts necessary to extract useful carriers from the comb source. Section 4.3 discusses two options to deal with the large optical linewidth. One is to reduce the optical linewidth in a feed-forward heterodyne phase noise reduction scheme, and the other one is to employ a phase-noise tolerant self-homodyne detection scheme. Using the first approach, transmission of 1 Tbit/s over 75 km with QPSK could be demonstrated [C6], and the latter approach allowed for the transmission of 1.5 Tbit/s over 75 km with 16QAM [C7]. The chapter is summarized and concluded in Section 4.4.

4.1 Device structure and optical spectrum

QD-MLLD were developed over the past decade by several groups. A good historical overview can be found in [63]. The devices used for the current work are Fabry-Perot lasers driven with a constant current. They are sometimes also referred to as self-pulsating lasers [140], [141] as the output signal is pulsed despite being driven with DC. The physical origin of the pulsed output can be understood as follows: Due to the inhomogeneously broadened gain, multiple modes are excited simultaneously, spaced by the free-spectral range (FSR) of

---

[9] http://www.bigpipes.rwth-aachen.de/

the laser cavity, which is defined by cleaved facets. Mode-locking is attributed to third-order optical nonlinear interactions inside the cavity [63]–[65].

Fig. 4.1 shows a simplified schematic of the cross-section of the device. The active region comprises several layers of InAs quantum-dashes separated by InGaAsP barriers in a dash-in-a-barrier design embedded in a separate confinement heterostructure. The electrical contacts are at top and bottom of the device. More details on the device structure and fabrication process can be found in [62].

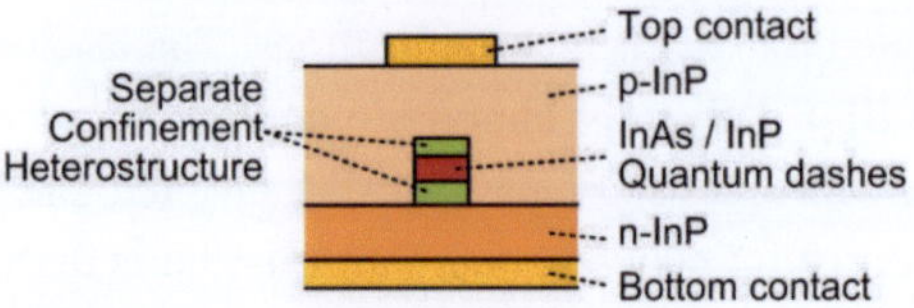

Fig. 4.1: Schematic cross-section of a quantum-dash mode-locked laser. The active region comprises several layers of InAs quantum dashes separated by InP barriers. It is embedded into a separate confinement heterostructure. Electrical contacts are placed at the top and bottom of the laser chip

A typical output spectrum of QD-MLLD with ~1 mm cavity length is depicted in Fig. 4.2. The line spacing corresponds to 42 GHz. A number of 40 lines are found which are flat within 3 dB. The total fiber-coupled output power amounts to 10 dBm. It is hence conceivable that this laser could replace as much as 40 individual lasers and their respective individual wavelength stabilization.

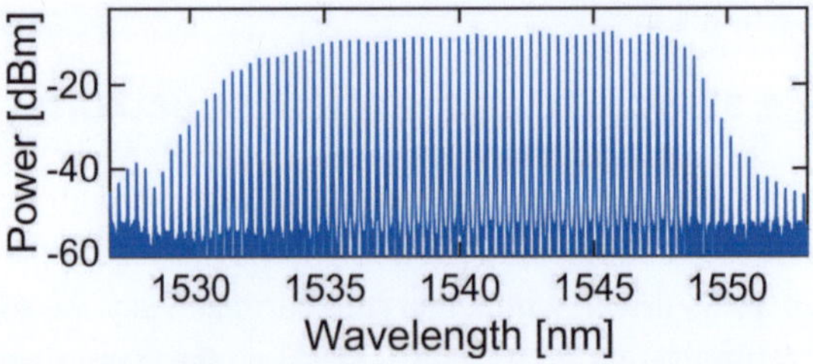

Fig. 4.2: Typical spectrum of a QD-MLLD. The laser is biased at 380 mA, the comb line spacing corresponds to 42 GHz, and the total fiber coupled output power amounts to 10 dBm. A number of 40 comb lines are found within a flatness of 3 dB.

4.2 Sample probing and pre-characterization

During our preliminary characterizations of the QD-MLLD we found that the laser performance heavily depends on the operation conditions and on environmental influences. This section discusses the major measures that were taken in order to obtain the best comb performance for the subsequent data transmission experiments. Furthermore, the techniques used to characterize the mode beating linewidth, the optical linewidth and the relative intensity noise are explained.

4.2.1 Probing station and isolation from environment

The QD-MLLD used in this work are single-section devices that do not have a saturable absorption region, see [63] for a comparison between single-section and two-section devices. Single-section devices are driven by a single DC current. The probing station that was designed to test the QD-MLLD comprises a sample holder with a rotary stage, a temperature sensor, and a Peltier element. As the laser is quite sensitive to temperature changes, we use a hiqh quality temperature controller[10]. The laser chip is placed on the sample holder, and two DC probe needles are used for current injection and grounding. A lensed fiber with a beam waist diameter of 4 μm is used to couple light from the laser into a single mode fiber. As the devices are very sensitive to optical feedback, we use lensed fibers with an anti-reflection (AR) coating. To obtain the best and reproducible performance, we characterize the return loss of the lensed fibers using an optical backscatter reflectometer[11] and use only fibers with a return loss of 50 dB or more. A microscope camera helps positioning the DC probe needles and the lensed fiber. After this alignment, the whole probing station is covered using a Plexiglas box in order to prevent airflow, which could cause displacement of the lensed fiber as well as temperature variations of the laser chip, both impeding stability of the laser.

4.2.2 Radio frequency beat map

A method to determine the optimum operating point of a QD-MLLD is to measure the radio frequency beat signal for different injection currents using an electrical spectrum analyzer (ESA). This measurement corresponds to detecting the entire comb spectrum with a high-speed photodetector. A signal

[10] http://www.thorlabs.de/thorproduct.cfm?partnumber=TED200C
[11] http://lunainc.com/obr4600

at the frequency corresponding to the FSR of the comb can be measured with an electrical spectrum analyzer. This signal corresponds to an overlay of the mixing product of each comb line with its respective neighbor. The result can be visualized in a map as depicted in Fig. 4.3 showing the frequency over current. The color-code shows the power spectral density (PSD) measured with the ESA. It can be clearly seen that the repetition rate of the comb is not a linear function of the current. There are several regions found, where the linewidth of the beat signal is either broad, consisting of multiple peaks, or small, see Fig. 4.3(b).

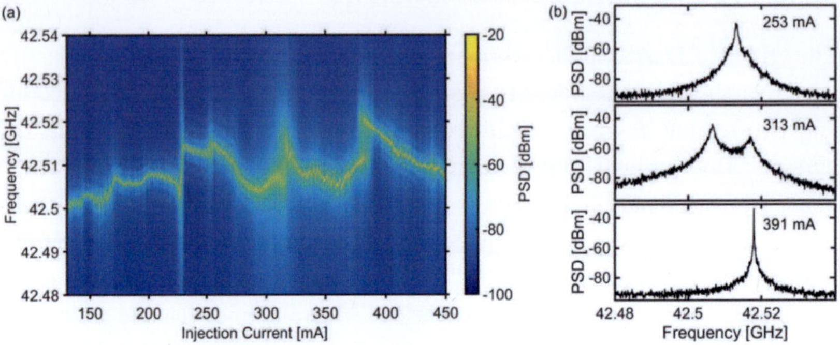

Fig. 4.3: Radio frequency beat map. (a) This map shows the evolution of the beat signal over current. For best performance the laser should be operated at a current where the linewidth of the beat signal is narrow. (b) Exemplary beat signals corresponding to vertical cross-sections of (a) showing from top to bottom a region with large linewidth, a region with multiple peaks and a region with narrow linewidth.

The occurrence of multiple peaks hints to the existence of super-modes [142], i.e. the co-existence of multiple frequency combs with slightly different line spacing. The optimum current ranges are characterized by a single and narrow peak, in this example the current range near 391 mA.

4.2.3 Optical linewidth

Narrow linewidth carriers are crucial for coherent data transmission as discussed in Section 2.2.2. Published measurements of optical linewidths are of the order of 10 MHz for comparable QD-MLLD, cf. [143], [144]. We confirm these findings by measuring the optical linewidth of the combs using the setup depicted in Fig. 4.4(a). A tunable external cavity laser (ECL) with a narrow linewidth, on the order of 10 kHz, is mixed with each comb line on a photodiode, generating a signal at the offset frequency between comb line and

ECL. The linewidth of this signal is measured using an electrical spectrum analyzer (ESA) and is assumed to be equal to the linewidth of the respective comb line. This assumption is justified as the ECL linewidth is about three orders of magnitude smaller and hence negligible in comparison to the measured linewidth. An optical spectrum analyzer is used to monitor the optical spectrum of the comb and the tuning of the ECL.

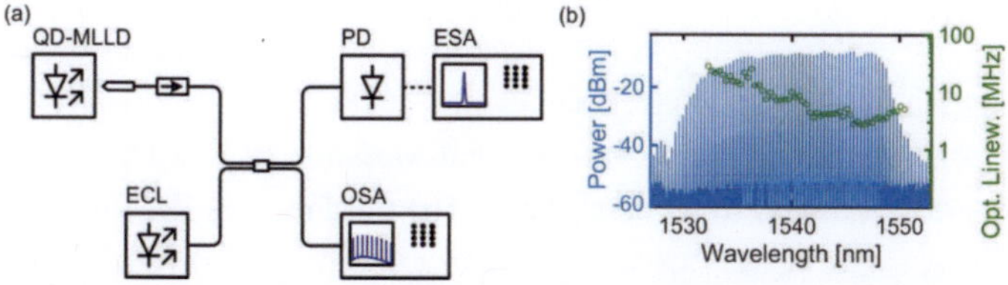

Fig. 4.4: Optical linewidth characterization. (a) Experimental setup. Each comb line of the QD-MLLD is mixed with an external cavity laser (ECL) by a photodiode (PD). The optical linewidth is determined by measuring the linewidth of the beat signal between comb line and the ECL. An optical spectrum analyzer monitors the comb spectrum. (b) Resulting measurement result.

An exemplary result is shown in Fig. 4.4(b). The plot shows the optical spectrum, left-hand axis, and the optical linewidth, right-hand axis, for each comb line. A minimum linewidth of 2.7 MHz is found, which increases to above 10 MHz for the outer lines. As discussed in Section 2.2.2, such wide optical linewidths introduce a strong penalty for coherent data transmission.

The linewidth of the comb lines is affected by several contributions [145]. First, there are phase fluctuations due to spontaneous emission. Those phase fluctuation represent the fundamental limit, the Schawlow-Townes limit. Second, spontaneous emission leads to amplitude noise, which in turn affects the carrier density of the laser diode and hence the refractive index of the cavity, which finally causes frequency noise, see also Appendix C.1. This effect is often taken into account using a linewidth enhancement factor, see Eq. (C.1). Third, timing jitter $\Delta\tau_r(t)$ of the intracavity pulses is responsible for an increase of the linewidth with the mode number.

The comb field E that is composed out of $2M+1$ modes can be expressed using the individual comb frequencies $f_m = f_0 + m \cdot f_r$ (Eq. (1.1)), their amplitudes A_m, and their phases [145], [C3]

$$\varphi_m(t) = \phi_m + \varphi_{\text{inc}}(t,\tau) + 2\pi f_r m \Delta\tau_r(t) \tag{3.1}$$

as

$$E = \sum_{m=-M}^{M} A_m \cos\left(2\pi f_m t + \varphi_m(t)\right). \tag{3.2}$$

The first term in Eq. (3.1) is the static phase ϕ_m of each comb line. The second term in Eq. (3.1) corresponds to the increments of the phase φ after an observation time τ

$$\varphi_{\text{inc}}(t,\tau) = \varphi(t-\tau) - \varphi(t). \tag{3.2}$$

The fluctuations of the phase increments belong to a stationary Gaussian process with diffusion coefficient D and variance $\sigma^2_{\varphi_{\text{inc}}} = D|\tau|$ [144]. This phase noise is common to all comb lines. The third term in Eq. (3.1) shows the contribution to the phase due to fluctuations $\Delta\tau_r(t)$ of the round-trip time of the intracavity pulses. As these phase fluctuations increase in proportion to m, the associated phase noise variances $\sigma^2_{\varphi_m}$ of the comb lines increase by m^2

$$\sigma^2_{\varphi_m} = \sigma^2_{\varphi_{\text{inc}}} + m^2\sigma^2_{\varphi_r}. \tag{3.3}$$

In the experiments described in Section 4.3.1, $\varphi_{\text{inc}}(t)$ is estimated using one of the comb lines. Afterwards, all comb lines are phase modulated with $-\varphi_{\text{inc}}(t)$ for phase noise compensation [146]. To estimate $\varphi_{\text{inc}}(t)$, the comb line with index s, having the field strength $E_s = A_s \cos\left(2\pi f_s t + \varphi_s(t)\right)$, is selected. A balanced receiver with sensitivity S mixes this comb line with a narrow-linewidth local oscillator $E_{\text{LO}} = A_{\text{LO}} \cos\left(2\pi f_{\text{LO}} t + \varphi_{\text{LO}}(t)\right)$. The intermediate frequency is defined as $f_z = f_{\text{LO}} - f_s$. An electrical bandpass filter selects the signal $z(t)$, which can be written as

$$z(t) = S A_s A_{\text{LO}} \cos\left(2\pi f_z t - \varphi_s(t) + \varphi_{\text{LO}}(t)\right). \tag{3.4}$$

All comb lines are modulated with $z(t)$ using a Mach-Zehnder modulator (MZM), operated in the linear regime. Considering the delay τ_{dly} between E and z, the output signal $E_c(t)$ becomes

$$
\begin{aligned}
E_c(t) &= \sum_{m=-M}^{M} A_m \cos\left(2\pi f_m\left(t-\tau_{\text{dly}}\right)+\varphi_m\left(t-\tau_{\text{dly}}\right)\right) \\
&\quad \times S A_s A_{\text{LO}} \cos\left(2\pi f_z t-\varphi_s(t)+\varphi_{\text{LO}}(t)\right) \\
&= \frac{1}{2} S A_s A_{\text{LO}} \sum_{m=-M}^{M} A_m \cos\left(2\pi\left(f_m-f_z\right)t-2\pi f_m \tau_{\text{dly}}\right. \\
&\qquad\qquad \left.+\varphi_m\left(t-\tau_{\text{dly}}\right)+\varphi_s(t)-\varphi_{\text{LO}}(t)\right) \\
&\quad +\frac{1}{2} S A_s A_{\text{LO}} \sum_{m=-M}^{M} A_m \cos\left(2\pi\left(f_m+f_z\right)t-2\pi f_m \tau_{\text{dly}}\right. \\
&\qquad\qquad \left.+\varphi_m\left(t-\tau_{\text{dly}}\right)-\varphi_s(t)+\varphi_{\text{LO}}(t)\right).
\end{aligned}
\tag{3.5}
$$

It can be seen that the modulation process generates two sidebands at $f_m \pm f_z$, while the original comb line is suppressed. Choosing a sufficiently small τ_{dly}, the phase noise is suppressed for the case $f_m + f_z$, while the phase noise doubles for the other case $f_m - f_z$. The phase noise compensated lines can be filtered and used as carriers for coherent data transmission, which is described in Section 4.3.1. To this end an intermediate frequency of $f_z = 8\,\text{GHz}$ is used. Note that alternatively it is possible to use an IQ modulator driven as a frequency shifter to obtain only the lines with reduced phase noise. This would remove the need for filtering the unwanted comb lines that have twice the phase noise.

4.2.4 Relative intensity noise

Relative intensity noise (RIN) of the individual comb lines is a very important parameter for data transmission as described in Section 2.2.1. In [147], RIN measurements for individual comb lines of a QD-MLLD as well as for the entire comb are compared. These measurements show that the RIN of all comb lines together is extremely small, however, the RIN of the individual comb lines is very high. This effect is commonly referred to as mode partition noise [148]. Physically, it means that the modes constantly exchange energy amongst each other.

In Section 2.2.1 we introduced the OCNR as a rough estimation of the RIN. However, this measurement neglects the frequency dependence of the intensity noise. In the experiment described in Section 4.3.2 a deformation of the constellation diagram is found for certain comb lines. It turns out that the comb lines that show this behavior feature a resonance in the $\mathrm{RIN}(f)$ measurement. Using the setup depicted in Fig. 4.5(a) we characterize the RIN of the comb lines that we want to use for data transmission. To this end we use a narrow programmable filter (Finisar WaveShaper) to select a single line. As the power of a single line at the output of the band-pass filter (BPF) is only of the order of −18 dBm, we amplify the line with an erbium-doped fiber amplifier (EDFA). An additional BPF is used to filter noise from the EDFA. The comb line is then detected using a photodiode (PD). A bias-T separates DC and AC part of the photocurrent. An ampere meter measures the DC part I_{PD}, while an ESA measures the power spectral density (PSD) of the photocurrent. The RIN and $\mathrm{RIN}(f)$ can then be calculated using Eqs. (C.2) and (C.9), from Appendix C.2.

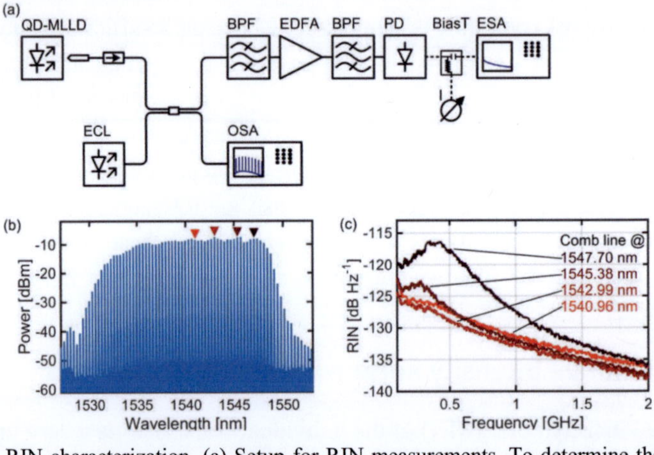

Fig. 4.5: RIN characterization. (a) Setup for RIN measurements. To determine the RIN of individual comb lines a narrow band-pass filter (BPF) selects a single comb line that is subsequently amplified using an erbium-doped fiber amplifier (EDFA). Noise from the EDFA is filtered using a second BPF before the comb line is detected by a photodiode (PD). A bias-T is used to separate the DC and AC part of the photocurrent. The DC part is measured using an ampere meter, while an electrical spectrum analyzer measures the power spectral density of the AC part in the range from 6 MHz to 2 GHz. An external cavity laser (ECL) with a low RIN (< -145 dB Hz^{-1}) is used to calibrate the measurement. (b) Spectrum of device, biased at 380 mA. (c) RIN results for the lines indicated by the triangles in (b).

Fig. 4.5(b) shows the spectrum of the laser, biased at 380 mA, and four triangles marking the comb lines for which RIN measurements are shown in Fig. 4.5(c). For this device we observe at low frequencies a RIN level on the order of −125 dB Hz^{-1} for most comb lines, similar as the two lines at

1540.96 nm and at 1542.99 nm shown in the figure. Lines at the far red edge of the spectrum feature a peaking of the RIN at around 400 MHz. This behavior has been observed for the first time in our measurements and its origin is yet unclear. The consequences for the performed data transmission experiment will be discussed in Section 4.3.2.

4.3 Data transmission experiments

This section discusses two possibilities to use QD-MLLD for coherent data transmission despite their large optical linewidth. The first approach is to implement a phase noise reduction scheme as introduced in Section 4.2.3. This reduces the phase noise of the comb lines directly at the transmitter side. The second approach is to change the system architecture to a self-homodyne detection scheme. This approach transmits unmodulated carriers along with the signal and uses them at the receiver site as local oscillator. Phase noise of the carriers and the local oscillators is then correlated and hence the phase error is reduced.

4.3.1 Managing phase noise at the source

This section corresponds to [C6].

Simultaneous Phase Noise Reduction of 30 Comb Lines from a Quantum-Dash Mode-Locked Laser Diode Enabling Coherent Tbit/s Data Transmission

Optical Fiber Communication Conference, *2015, paper Tu3I.5.*

J. Pfeifle[1], R. Watts[2], I. Shkarban[1,3], S. Wolf[1], V. Vujicic[2], P. Landais[2], N. Chimot[4], S. Joshi[4], K. Merghem[5], C. Calò[5], M. Weber[3], A. Ramdane[5], F. Lelarge[4], L. Barry[2], W. Freude[1,6], C. Koos[1,6]

[1]*Institute of Photonics and Quantum Electronics (IPQ), Karlsruhe Institute of Technology (KIT), Germany*

[2]*The Rince Institute, School of Electronic Engineering, Dublin City University, Glasnevin, Dublin 9, Ireland*

[3]*Institute for Data Processing and Electronics (IPE), Karlsruhe Institute of Technology (KIT), Germany*

[4]*III-V Lab, Marcoussis, France*

[5]*Laboratoire de Photonique et Nanostructures, CNRS UPR20, Marcoussis, France*

[6]*Institute of Microstructure Technology (IMT), Karlsruhe Institute of Technology (KIT), Germany*

A feed-forward heterodyne scheme is shown to simultaneously reduce the phase noise of many comb lines from a quantum-dash mode-locked laser diode (QD-MLLD). This enables the first coherent data transmission using QD-MLLD as multi-wavelength source.

Introduction

Over the past years several experiments have demonstrated the huge potential of using optical frequency combs for data transmission at Tbit/s rates [51], [53], [J4], [J5]. For practical systems, a compact, low-cost, energy-efficient comb source is desired. Single-section quantum-dash mode-locked laser diodes (QD-MLLD) have been shown to deliver several tens of comb lines with decent optical power while being driven simply by a constant current of the order of 100 mA [65]. However, the spectral modes of QD-MLLD generally exhibit strong phase noise and broad optical linewidths, typically on the order of 10 MHz [143], [149]. As a consequence, wavelength-division multiplexing (WDM) data transmission using these combs has been restricted to direct detection schemes [150] and differential quadrature phase shift keying (DQPSK) with high symbol rates and aggregate data rates of up to 504 Gbit/s [147], whereas fully coherent transmission has been impeded by excessive phase noise up to now. Optical injection using a narrow-linewidth master laser can reduce the optical linewidth of QD-MLLD, but also distorts the spectral envelope of the comb and significantly reduces the available bandwidth [151], thereby limiting its utility for broadband WDM systems. Digital feed-forward schemes have been proposed to reduce the influence of excessive phase noise in either the transmitter laser or local oscillator (LO) laser in coherent receivers [152], [153], and these digital schemes could be employed to permit using QD-MLLD in coherent systems. Likewise, an analogue equivalent has recently been shown to reduce the phase noise for multiple comb lines simultaneously by using a feed-forward heterodyne (FFH) scheme [146]. However, the viability of this concept for high-speed coherent data transmission has not yet been shown.

In this work, the FFH scheme [146] is demonstrated to simultaneously reduce the linewidth of 30 lines from a QD-MLLD with 42 GHz line spacing. The resulting lines are used for coherent WDM data transmission at Tbit/s rates. This approach has, for the first time, permitted coherent transmission with a semiconductor QD-MLLD.

FFH phase noise reduction for a 42 GHz QD-MLLD

The QD-MLLD used in this work has a free spectral range of 42 GHz. The active region consists of 6 layers of InAs quantum dashes emitting at 1.55 μm, separated by InGaAsP barriers in a dash-in-a-barrier design [62]. A buried ridge waveguide of 1.25 μm width confines the optical mode. Feedback is provided by the cleaved end facets of the QD-MLLD. The total length of the device is 980 μm. We operate the QD-MLLD at a constant current of 338 mA, which is optimized for a narrow RF beatnote featuring 16 kHz full width at half maximum (FWHM) along with minimal side peaks. The combination of large optical linewidth and a narrowband RF beatnote indicates a high correlation of the comb line's phases [142]. The output spectrum is depicted in Fig. 4.6(a).

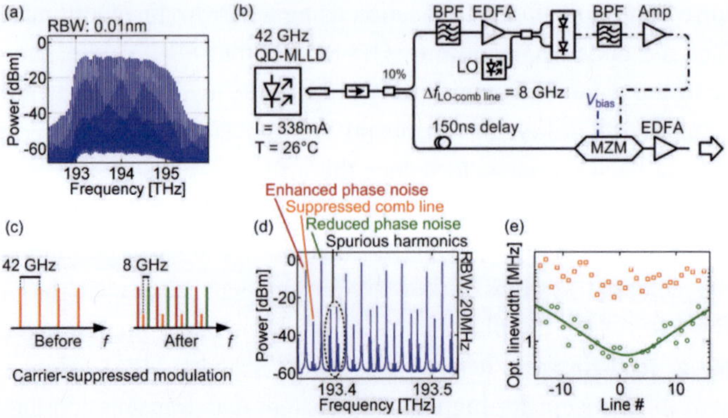

Fig. 4.6: Setup for feed-forward heterodyne (FFH) phase noise reduction. (a) Optical spectrum of the 42 GHz comb featuring a 3 dB bandwidth close to 1.4 THz and an average output power of 10 dBm. (b) Experimental setup. Light from the QD-MLLD is collected using an anti-reflection coated lensed fiber followed by an isolator to avoid backreflection into the laser cavity. For the FFH scheme, a small portion of the comb is tapped and a single line is filtered from the comb spectrum. This line is mixed with a narrow-linewidth LO laser that is offset by 8 GHz. The resulting RF beat generated in a balanced photodetector is filtered, amplified and fed to a MZM that is biased at minimum transmission for carrier-suppressed modulation. This modulation generates two sidebands around each comb line, cf. (c). The upper sidemode (green line) exhibits reduced phase noise while the lower sidemode (red line) features enhanced phase noise. (d) Optical spectrum of the output from the FFH-scheme showing pairs of lines surrounding the suppressed comb line. The two smaller peaks within the free spectral range correspond to spurious harmonics generated by the carrier-suppressed modulation. (e) Result of the heterodyne optical linewidth measurement for the original comb lines (orange) and the lines with reduced phase-noise at the output of the FFH-scheme (green). It can be clearly seen that the linewidth of all modes is reduced. Line # 0 corresponds to the comb line that is mixed with the offset LO and is used to generate the MZM drive signal.

The FFH phase noise reduction scheme is illustrated in Fig. 4.6(b). A small percentage of the power of the comb is tapped, a single line is filtered out, amplified and mixed with a narrow-linewidth local oscillator (LO) laser (Koheras AdjustiK) that is tuned to an offset of 8 GHz from the comb line. The beat signal at 8 GHz, created by detection with a balanced photodetector, is filtered, amplified and fed to a Mach-Zehnder modulator (MZM) that is biased at zero transmission and driven by a sinusoidal. The MZM generates two sidemodes at ±8 GHz while suppressing the original comb line. It was shown in [146] that the upper sidemode (green lines in Fig. 4.6(c) and (d)) exhibits reduced phase noise. Other offset frequencies could also chosen – for our setup, 8 GHz represents the best tradeoff between a high offset to allow separating the two sidebands and a low offset to not run into bandwidth limitations of the equipment.

We confirm the optical linewidth reduction through a direct measurement, see Fig. 4.6(e). To this end we beat each line with a narrow-linewidth laser and evaluate the width of the beat measured with a fast photodiode and a radio frequency (RF) spectrum analyzer [154]. The FFH scheme reduces the optical linewidth for all 30 modes that have been tested. For the center modes, the resulting linewidth is below 1 MHz, which is sufficiently low for application in coherent data transmission with current DSP. Typical linewidth requirements for an 18 GBd QPSK signal are 3.8 MHz [118] or 7.4 MHz [119], depending on the receiver implementation.

Coherent Tbit/s data transmission employing FFH phase-noise reduction

After FFH phase noise reduction, the frequency comb exhibits pairs of lines that correspond to the upper and lower modulation sidemode, Fig. 4.6(c). For data transmission we want to use the upper sidemodes (green lines) which feature the reduced phase noise. Fig. 4.7(a) shows a schematic of the WDM data transmission testbed used in this work. The upper sidemodes are filtered and dis-interleaved into odd and even carriers using a Finisar waveshaper (WS) while the lower sidemodes are suppressed. After dis-interleaving, we modulate even and odd carriers with 18 GBd sinc-shaped QPSK signals using a pseudo-random bit sequence of length 2^{11}-1. The signals are combined and either tested in (A) a back-to-back (B2B) configuration, or (B) amplified and sent through a 75 km spool of standard single mode fiber (SSMF). The receiver comprises filters, a pre-amplifier and an optical modulation analyzer (OMA, Keysight N4391A) that uses coherent detection with an external cavity laser (ECL) as LO. We evaluate the signal quality using the Keysight VSA software. Signal processing includes digital brick-wall filtering, dispersion compensation, and adaptive equalization before evaluating the bit-error-ratio (BER). Fig. 4.7(b) shows constellation diagrams with and without using the FFH scheme for the case of line # 0. The reduction of the phase error can clearly be seen.

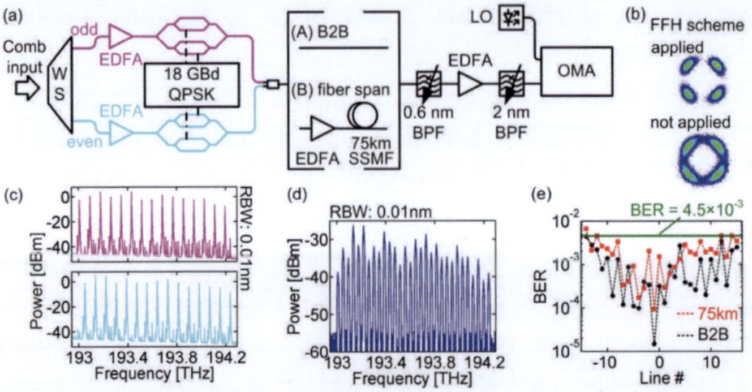

Fig. 4.7: Setup for WDM data transmission and measured results. (a) Upper sidemodes of the output from the FFH scheme are selected and split into odd and even carriers for independent modulation with QPSK signals at a symbol rate of 18 GBd. The signals are combined using a 3 dB coupler and tested either in a back-to-back (B2B) configuration or amplified and transmitted over 75 km of standard single-mode fiber (SSMF). The receiver comprises filters, a pre-amplifier and an optical modulation analyzer (OMA) that uses a coherent detection scheme with an ECL as local oscillator (LO). (b) Constellation diagrams obtained with and without the FFH-scheme. (c) Spectra of odd (top) and even (bottom) comb lines after dis-interleaving. (d) Optical spectrum of the 30 data channels. (e) Bit error ratio results for 30 channels when applying the FFH-scheme. An aggregate data rate of 0.971 Tbit/s could be recovered within the limits of forward-error correction with 7 % overhead.

Fig. 4.7(c) shows the dis-interleaved comb spectra. The extinction ratio of the lower, unwanted sidemode is better than 26.5 dB with an average of 33.8 dB. This was achieved by narrowing the bandwidth of the programmed filters, which unfortunately also leads to an additional attenuation for some of the carriers due to the inherent interpolation of the filter shape in the WS, where each pixel covers a frequency range of about 5 GHz. Note that this only represents a limitation of our measurement setup and could be solved when using, e. g., a periodic filter with narrow passbands such as a high-finesse Fabry-Perot etalon. A second possibility to increase the extinction ratio would be to replace the MZM in the FFH scheme with an IQ modulator driven as a frequency shifter, thereby only generating a single sidemode. The spectrum of the modulated comb is depicted in Fig. 4.7(d) for the B2B case. It comprises 30 channels, which corresponds to a line rate of 1.08 Tbit/s. The average OSNR of the channels at the Rx (input to the 0.6nm filter) is 26.4 dB for the B2B case and 25.0 dB after 75 km fiber transmission. BER results for B2B and 75 km are depicted in Fig. 4.7(e). For B2B, all 30 channels are below the BER threshold of 4.5×10^{-3} for a standard forward-error correction (FEC) scheme with 7 % overhead, corresponding to a net data rate of 1 Tbit/s. After transmission, the channel with index –14 exceeds the FEC limit, resulting in only 29 channels within the 7 % FEC limit. This corresponds to a net data rate of 0.971 Tbit/s after transmission. The data rate could be doubled with polarization multiplexing, which could not be tested in these experiments due to limited availability of hardware. Nevertheless, our results demonstrate that FFH phase-noise reduction can be used to simultaneously reduce the optical linewidth of a multitude of optical lines that are derived from one semiconductor QD-MLLD, thereby enabling the first coherent data transmission experiment using a frequency comb from a QD-MLLD.

Summary

We demonstrate a FFH scheme that enables simultaneous optical linewidth reduction of a multitude of comb lines derived from a QD-MLLD. The scheme enables the first coherent transmission experiment using QD-MLLD as an optical source. Net data rates of nearly 1 Tbit/s could be achieved within the limits of standard FEC.

4.3.2 Managing phase noise by system design

This section is published in [C7].

Coherent Terabit Communications Using a Quantum-Dash Mode-Locked Laser and Self-Homodyne Detection

Optical Fiber Communication Conference, *2015, paper W2A.19.*

J. Pfeifle[1], I. Shkarban[1,2], S. Wolf[1], J. N. Kemal[1], C. Weimann[1], W. Hartmann[1,3], N. Chimot[4], S. Joshi[4], K. Merghem[5], A. Martinez[5], M. Weber[2], A. Ramdane[5], F. Lelarge[4], W. Freude[1,3], C. Koos[1,3]

[1]*Institute of Photonics and Quantum Electronics (IPQ), Karlsruhe Institute of Technology (KIT), Germany*

[2]*Institute for Data Processing and Electronics (IPE), Karlsruhe Institute of Technology (KIT), Germany*

[3]*Institute of Microstructure Technology (IMT), Karlsruhe Institute of Technology (KIT), Germany*

[4]*III-V Lab, Marcoussis, France*

[5]*Laboratoire de Photonique et Nanostructures, CNRS UPR20, Marcoussis, France*

We transmit 18 GBd 16QAM signals on 25 spectral lines of a quantum-dash mode-locked laser diode, achieving a 1.562 Tbit/s aggregate data rate. Phase noise is cancelled by self-homodyne detection using LO tones transmitted with the signal.

Introduction

Frequency combs are promising multi-carrier sources for future point-to-point data links operating at Tbit/s data rates [51], [53], [J4], [J5]. Single-section quantum-dash mode locked laser-diodes (QD-MLLD) are very compact comb sources that exhibit broadband spectra comprising many lines with a flat power distribution while being driven at a constant current [65]. A drawback of the QD-MLLD is the large optical linewidth of the comb lines, which is typically of the order of several tens of MHz [143], [149]. As a consequence, data transmission experiments with these sources were so far restricted to on-off-keying [150] or to differential quadrature phase shift (DQPSK) keying [147], where the highest demonstrated data rate was 504 Gbit/s. For using a QD-MLLD comb to coherently transmit data it would be required to reduce the optical linewidth of the comb lines. This was demonstrated using a feed-forward heterodyne phase noise reduction scheme [146], however, the power per comb line was greatly reduced as well and the linewidth reduction is only efficient for the central modes.

In this paper, we demonstrate the first coherent data transmission of a 16QAM signal using a QD-MLLD and employing self-homodyne detection. To this end, a comb of LO tones derived from the same comb source used to transmit the signal [155] is sent along with the signal. Using 25 comb lines, we achieve an aggregate net data rate of more than 1.5 Tbit/s after transmission over 75 km of standard single-mode fiber. The line rate per wavelength channel amounts to 72 Gbit/s, corresponding to a single-polarization symbol rate of 18 GBd for a 16QAM format. We show that relative intensity noise of isolated QD-MLLD comb lines is largest for lines at the red edge of the spectrum, and we observe a correlation between amplitude and phase noise for these lines.

Experimental setup for comb-based WDM data transmission using a self-homodyne coherent system

With self-homodyne coherent systems, an unmodulated carrier is transmitted as a local oscillator (LO) along with the modulated signal. For separating the

LO from the signal at the receiver side, either a spatial or polarization multiplexing technique is used [155].

The experimental setup used in this work is depicted in Fig. 4.8. The comb spectrum is split into two paths, indicated in green and blue in Fig. 4.8. The upper path (green) emulates a wavelength-division multiplexing (WDM) transmitter by splitting the comb lines into odd and even carriers, and by modulating independent data from a pseudo-random bit sequence of length 2^{11}-1 on the two sets of carriers. As modulation format we use 16QAM in combination with sinc-pulse shaping at a symbol rate of 18 GBd. Odd and even data channels are merged using a 3 dB coupler before the data stream is recombined with the unmodulated comb lines on orthogonal polarizations of a standard single-mode fiber (SSMF). The receiver comprises a polarization controller and polarization beam splitter (PBS) to separate the channels from the LO lines. Filters in both paths select the respective channel and the corresponding LO. For detection and signal analysis we use an optical modulation analyzer (Keysight N4391A OMA). For efficient cancellation of the phase noise, the group delay for both signal and LO need to be matched. To this end we fit the fiber lengths of the green and blue paths in Fig. 4.8 and use a tunable delay line for fine adjustment. The tolerable path length difference depends on the linewidth of the laser and on the tolerable phase error [155]. For a 10 MHz linewidth and maximum phase error of 5° we estimate a maximum tolerable path length difference of 30 cm. This corresponds to a differential group delay of approximately 0.7 ns. Comparing this value to the polarization mode dispersion (PMD) of SSMF, which is typically well below $1\ \mathrm{ps}/\sqrt{\mathrm{km}}$, it is not to be expected that PMD is a limiting factor for the phase noise cancellation. The signal quality is characterized, first in a back-to-back (B2B) configuration and second after transmission over a 75 km transmission span, by measuring the bit-error-ratio (BER).

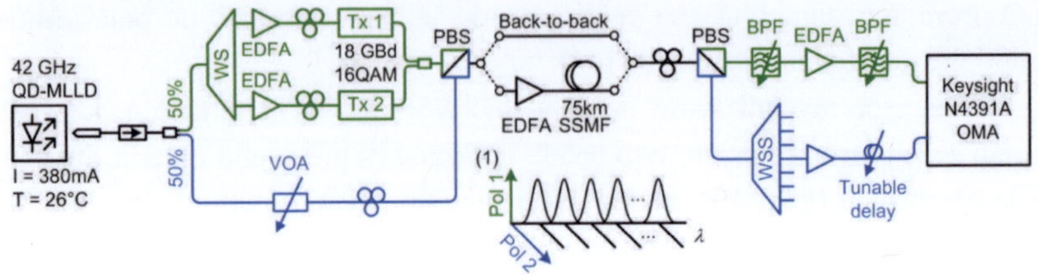

Fig. 4.8: Experimental setup for self-homodyne data transmission. On the transmitter side, a frequency comb is generated from a single-section QD-MLLD. Light is collected via an anti-reflection coated lensed fiber and passed through an isolator. For self-homodyne data transmission, the comb spectrum is split in two paths (green and blue). In one path (green) the comb is dis-interleaved into odd and even comb lines for independent modulation with data from a PRBS of length 2^{11}-1 using 16QAM at a symbol rate of 18 GBd along with sinc-pulse shaping. The other path (blue) bypasses the modulator and represents the local oscillator (LO) comb. Both paths are recombined for transmission on orthogonal polarizations using a polarization beam splitter (PBS), see Inset (1). Multi-channel signal and LO comb are either tested in a back-to-back (B2B) configuration or transmitted over 75 km of standard single-mode fiber (SSMF). At the receiver's side, a second PBS separates the signal and the LO comb. The respective channel under test as well as the corresponding LO line are selected with filters (BPF – band-pass filter, WSS – wavelength selective switch), and detected using an optical modulation analyzer (OMA). The path lengths of signal (green) and LO (blue) need to be well matched so that the phase noise of signal and LO are correlated when received by the OMA. Fine-tuning is performed using the tunable delay line in the LO path.

Comb and back-to-back signal characterization

The laser chip is temperature-controlled at 26°C and contacted with DC probes for injecting a constant current of 380 mA. Light is coupled using a lensed fiber which collects an output power of 10 dBm. An anti-reflection coating on the fiber tip as well as a fiber isolator prevent backreflection from the transmission setup into the laser. The laser has a total length of 980 µm. Its active region consists of 6 layers of InAs quantum-dashes emitting at 1.55 µm, separated by InGaAsP barriers in a dash-in-barrier design [62]. The optical mode is confined using a buried ridge waveguide with a ridge width of 1 µm. Fig. 4.9(a) shows the spectrum (RBW 0.01 nm) of the comb comprising 40 lines within a flatness of 3 dB.

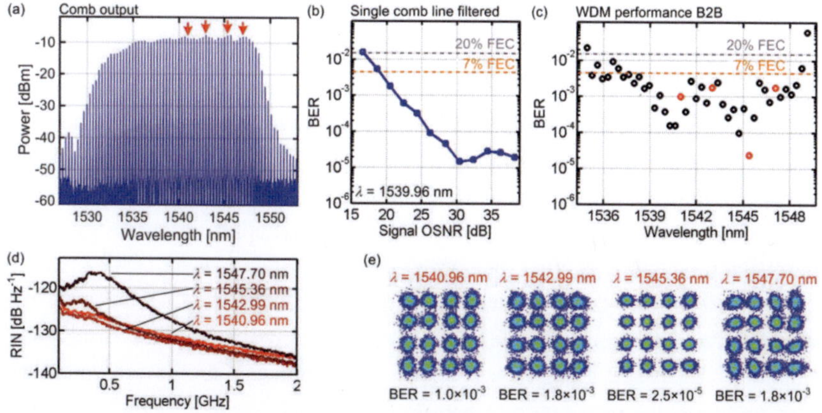

Fig. 4.9: Comb and signal quality characterization. (a) Comb spectrum measured with a resolution bandwidth of 0.01 nm when driving the MLLD at a constant current of 380mA. The comb spectrum comprises 40 lines within a spectral flatness of 3 dB, featuring an optical carrier-to-noise-density power ratio (OCN_0R) beyond 136 dB Hz. The total output power of the comb measured after the isolator is 10 dBm. (b) The bit error ratio (BER) vs. OSNR (signal and noise power within 19 GHz) when filtering a single comb line at the transmitter side shows the performance of the experimental setup. The signal is modulated at 18 GBd using 16QAM in combination with sinc pulse-shaping. BER limits for two different forward-error correction implementations are indicated. (c) BER results for 43 WDM channels in back-to-back (B2B) configuration. A number of 37 channels performs better than the 7 % FEC limit, and 4 channels are still below the limit for 20 % FEC. This corresponds to a total net data rate of 2.678 Tbit/s. (d) Relative intensity noise (RIN) of four selected comb lines, as indicated by the red arrows in (a) and red circles in (c). (e) Constellation diagrams derived from the same four comb lines. The constellation diagram for the channel at 1547.7 nm shows severe distortions along with a strong correlation of amplitude and phase noise. This is attributed to the higher RIN of this comb line and amplitude-phase coupling inside the laser cavity.

For a preliminary system characterization, we filter and modulate a single comb line near the center of the spectrum. Fig. 4.9(b) shows the bit-error ratio (BER) of this channel versus the OSNR for the back-to-back case. The OSNR relates signal and noise power in a bandwidth of 19 GHz, which comprises the total signal power. We change the OSNR by attenuating the filtered comb line with the WS and operating the following EDFA in constant output power mode. The two horizontal lines indicate the BER limits for two different implementations of forward-error correction (FEC) featuring an overhead of 7 % and 20 % [138], [139]. The error floor is attributed to hardware limitation of our transmitter and receiver setup. BER results using the WDM transmitter as shown in Fig. 4.8 are presented in Fig. 4.9(c) for a total of 43 channels, ranging from 1535.27 nm to 1548.42 nm. For channels outside this wavelength range, the signal quality was too low. For a total of 37 (4) channels we achieve a BER below the FEC limit with 7 % (20 %) overhead, which corresponds to an aggregate net data rate of 2.678 Tbit/s. Four channels are highlighted in red. For each of these channels, we measured the spectrum of the relative intensity noise (RIN) and the corresponding constellation diagram, see Fig. 4.9(d) and (e). The RIN of the comb line at 1547.7 nm is greatly enhanced and peaks at a frequency of roughly 400 MHz. The constellation diagram of the corresponding channel does not show pure amplitude noise, which would mean symbols elongated in radial direction, but a rather odd rotation of the elongated noise cloud. This hints at a correlation between amplitude and phase noise. We measured a correlation coefficient ρ between the deviations of the amplitude and the phase with respect to the ideal constellation points of about 1000 symbols and found a correlation in the order of $\rho = 0.2$ for this case. For constellations without this anomaly we measured a correlation of only $\rho \leq 0.05$. We attribute this to wavelength-dependent amplitude-phase coupling inside the laser cavity. Due to inhomogeneous broadening in the quantum-dash gain material, this effect influences only the respective comb line, and has little to no effect for other comb lines. The distortions of the constellations as well as the increased RIN are only observed for a few comb lines surrounding the wavelength range at 1547.7 nm. To the best of our knowledge, this behavior of a QD-MLLD comb has not been reported before and is subject to further investigations.

16QAM transmission on 25 comb lines over 75 km SSMF

Lastly we transmit the data stream over a span of 75 km standard single-mode fiber (SSMF). This requires amplification by an EDFA to overcome the loss of the fiber as well as additional dispersion compensation by digital signal processing at the receiver. For this transmission experiment we cannot use the full power of the unmodulated LO lines as that would saturate the in-line EDFA and hence also limit the gain for the signal. We optimize the system by attenuating the unmodulated comb lines until we find the best BER performance. The input powers of the in-line EDFA correspond then to -9.4 dBm and 1.7 dBm for the channels and for the unmodulated LO comb lines, respectively, and the EDFA output power is fixed at 19 dBm. Fig. 4.10 shows the measured BER for all channels that had a sufficient quality for measuring the BER. We find 13 (12) channels with a BER below FEC limit with 7 % (20 %) overhead. This corresponds to an aggregate net data rate of

1.562 Tbit/s. In our experiment, we employed sinc-pulse shaping, which would allow to further increase the spectral efficiency by either modulating at a higher symbol rate or by using a comb with smaller free spectral range.

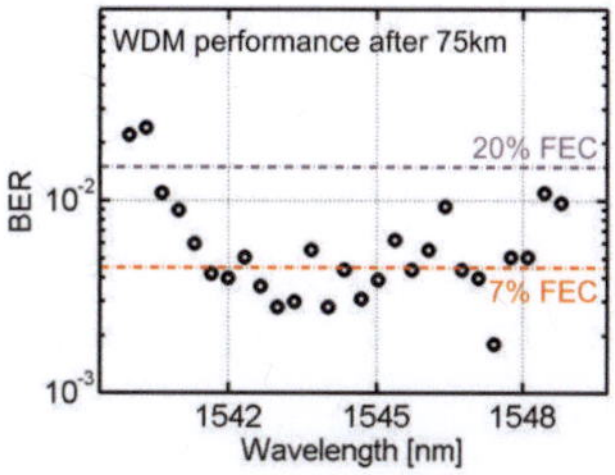

Fig. 4.10: Data transmission results after 75 km of SMF: BER vs. wavelength for 27 WDM channels. 13 (12) channels perform better than the BER threshold for FEC with 7 % (20 %) overhead. This corresponds to a net data rate of 1.562 Tbit/s.

Summary

We demonstrated the first 16QAM data transmission on comb lines from a QD-MLLD by utilizing a self-homodyne coherent system. For comb lines near the red side of the spectrum we observe a correlation of amplitude and phase noise that might be linked to a stronger RIN of these lines along with amplitude-phase coupling inside the laser cavity. In a WDM transmission scheme with 25 comb lines we achieved a post-FEC aggregate data rate of 1.562 Tbit/s.

4.4 Summary and conclusion

The advantages of the QD-MLLD are that they emit a broadband comb, spanning more than 10 nm optical bandwidth, while only requiring a single DC current supply. Furthermore, the spectral envelope of the comb is nearly rectangular such that only little power is lost in comb lines that cannot be used for data transmission. The compact size and simple operation makes these devices ready for system integration. In the experiments discussed in this chapter it is demonstrated that the QD-MLLD can be used for more spectrally efficient higher-order modulation formats like QPSK and 16QAM, and for bridging distances of the order of 75 km. Thus, QD-MLLD are promising candidates for data center interconnects and metro applications.

The data rates in both experiments can be further increased: During the experiments with the feed-forward phase noise compensation it was not possible to test polarization multiplexing due to lack of measurement equipment. Polarization multiplexing potentially increases the data rate by a factor of two. In addition, the 42 GHz free spectral range can much better be utilized by increasing the symbol rate from 18 GBd to 40 GBd. Such a high symbol rate is feasible with the next generation electronics. This would increase the total data rate by more than a factor of two. The same holds also for the data transmission experiments with the self-homodyne system architecture.

5 Kerr frequency comb generation for coherent Tbit/s communications

The generation of Kerr combs in non-linear microresonators is a promising technique as it allows to generate very broadband frequency combs spanning tens of THz with line spacings on the order of tens to hundreds of GHz. In principle, Kerr frequency comb sources solely consist of a single continuous wave pump laser and a microresonator. This makes it a very favorable candidate for chip-scale integration. It has been found that the properties of the frequency comb depend mostly on the dispersion profile of the resonator as well as on the applied pumping scheme. This also strongly affects the usefulness of the individual comb states for data transmission.

This chapter is structured as follows: The material platforms, resonator geometries, and comb formation dynamics are discussed in Section 5.1. Several data transmission experiments were performed using different Kerr comb sources. In Section 5.2 these experiments are discussed in detail, and Section 5.3 summarizes and concludes this chapter.

5.1 Formation of frequency combs in microresonators

This section provides the basic principles and concepts for understanding Kerr comb formation in microresonators. It will give a summary of experimental demonstrations of Kerr frequency combs and the associated resonator geometries and material systems that have been used. The concepts of primary combs, sub-combs and how this can lead to the formation of multiplet lines within the resonances will be introduced.

5.1.1 Material platforms and resonator geometries

Kerr comb generation has been demonstrated using various different material platforms and technologies to fabricate the required high-Q microresonator. Note that in the context of microresonators, Q is used as the quality factor of a resonator. This is not to be confused with the Q-factor metric that is introduced in Appendix B.4 in the context of eye diagrams to evaluate signal quality. The first demonstration of Kerr comb generation by Haye et al. dates back to 2007 [66] and uses a silica toroid. This resonator is fabricated by underetching a silica disc on a silicon substrate and by a subsequent reflow process that forms the toroidal ring standing on the silicon pillar. The reflow process ensures a smooth surface to achieve a high-Q whispering gallery mode[12] (WGM) resonator.

The next generation of Kerr comb sources were realized by milling and polishing a WGM resonator into a rod made from crystalline material as CaF_2 or MgF_2 [156], [157]. This is still a widely used technique as it allows to build large microresonators with FSR in the GHz-range and with very high Q-factors. The resonator dispersion can be engineered by adapting the geometry of the rim. Such a WGM resonator is used for the data transmission experiments that are discussed in Section 5.2.2.

A big milestone was achieved in 2010 by monolithic integration of a high-Q silicon-nitride (SiN) microresonator on a silicon substrate [74]. This opens up the route towards mass-producible Kerr comb generators that are compatible with CMOS technology to co-integrate other photonic and electronic components. SiN-based microresonators fabricated by the group of Tobias Kippenberg at École Polytechnique Fédérale de Lausanne (EPFL) are used for the experiments presented in Sections 5.2.1, 5.2.3, and 5.2.4.

Besides the aforementioned techniques, Kerr comb generation has also been demonstrated in a fiber Fabry-Perot resonator [158] and a ring resonator made from high-index glass [159]. A good summary of the early works and the applications of Kerr frequency combs can be found in [72]. Since then a couple of new developments were published, namely the wedge resonator using silica [160], and planar integrated ring resonators using single crystal diamond on silicon [161] and aluminum-nitride [162] for the waveguide layer. The latter is particularly suited for comb generation in the mid-infrared wavelength region, which is commonly used for spectroscopy [163]. Also silicon resonators have been proposed for this purpose [164], [165].

---

[12] A whispering gallery mode resonator is a resonator, where light is guided only at the outer interface between the dielectric, in this case air. That means that total internal reflection occurs only at this interface guiding the optical wave along the perimeter of the ring. The name is derived from acoustic whispering galleries. A famous example is St. Paul's Cathedral in London. Standing near the smooth wall of the circular gallery it is possible to understand the whisper of somebody standing near the same wall at any other position of the gallery.

5.1.2 Comb formation dynamics

To understand the difficulties and limitations when using a Kerr comb for a data transmission experiment, it is necessary to take a deeper look into the dynamics of the comb formation process. The explanations in this section follow [68], where the concept of primary combs, sub-combs and multiplet lines was first introduced and explained.

In Fig. 5.1(a) the initial situation of a resonator pumped with a strong continuous wave (CW) laser is depicted. The resonances (grey peaks) are non-equidistant due to dispersion leading to a frequency dependent FSR, cf. Appendix A.1. The CW pump is marked in red, and the blue areas indicate parametric gain due to FWM. The spectral separation between the parametric gain region and the pump depends on the dispersion of the cavity as well as on the pump power, see Appendix A.2 for a physical explanation. If the gain exceeds the loss of the cavity, new lines emerge at a distance Δ from the pump. Cascaded FWM processes lead to additional comb lines at integer multiples of Δ, forming a so-called primary comb. Such a primary comb shows high coherence and low noise. It is hence well suited for coherent data transmission as discussed in Section 5.2.2. However, the number of comb lines is strongly limited, as the power of the lines drops exponentially towards outer comb lines.

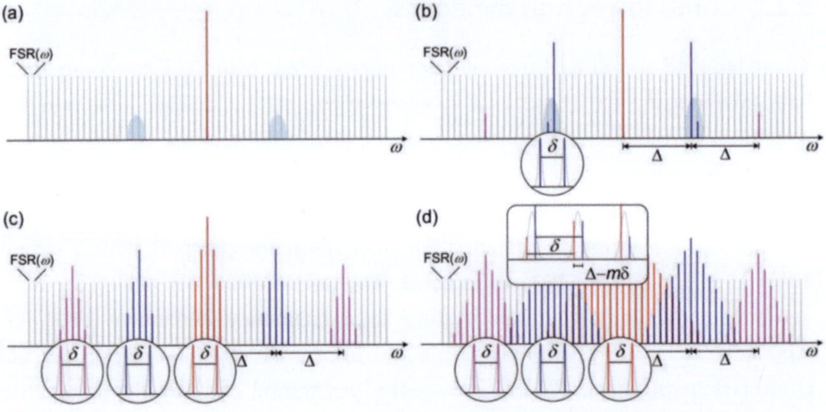

Fig. 5.1: Comb formation dynamics adapted from [68]. (a) Resonances with frequency dependent FSR. The red line marks the pump, and the blue shaded areas represent parametric gain due to FWM. (b) Primary comb lines emerge at a distance from the pump corresponding to integer multiples of Δ. Secondary comb lines with spacing δ form around the first primary comb lines. Generally Δ does not correspond to an integer multiple of δ. (c) The spacing δ is inherited by the new sub-combs forming around all primary comb lines as well as around the pump. (d) Overlapping of the sub-combs which are slightly offset with respect to each other, hence multiplet lines are found within single resonances as highlighted in the inset.

Fig. 5.1(b) and (c) illustrate the formation of sub-combs around the primary comb lines. At first, new lines emerge at the local resonance spacing δ. This can either be directly next to a pair of primary comb lines as depicted in Fig. 5.1(b), or right in between primary comb lines as discussed in [68]. All sub-combs inherit the same spacing δ. This can be understood when switching from a frequency-domain to a time-domain picture. In time-domain, FWM can be interpreted as a time-dependent modulation of the effective refractive index, which affects all modes in the same way. Once the frequency spacing δ is generated, it is therefore transferred across the whole comb spectrum [68]. Eventually, the individual sub-combs overlap forming a seamless comb spectrum, see Fig. 5.1(d). If the original primary comb spacing Δ differs from an integer multiple of δ, which is generally the case, then the overlapping sub-combs form multiplet lines within each resonance. This frequency mismatch has been observed to be in the MHz range [68]. Data transmission experiments are impeded by these multiplet lines. Particularly coherent communication is impossible with these combs. Also data transmission using intensity-modulation and direct-detection suffers from multiplet lines which cause a severe eye closure especially if there are two or more multiplets with similar power, see Section 5.2.1 for further details.

To avoid formation of multiplet lines it has been shown in [68] that it is possible to fine-tune the pump parameters such that the individual sub-combs oscillate on a common grid and thus add up coherently. Once the Kerr comb enters this synchronized state, it remains stable against fluctuations of the pump parameters [68]. The associated pumping scheme is sometimes referred to as "Δ–δ-matching". These comb states exhibit low phase noise and are very well suited for coherent data transmission. In Section 5.2.3, the experimental realization of such low phase-noise comb states and coherent Tbit/s data transmission are discussed.

5.2 Data transmission experiments

Using Kerr combs generated with the methods explained in the previous sections, several data transmission experiments were performed. The first generation of SiN microresonators that was supplied from the group of Tobias Kippenberg at EPFL[13] has been tested for WDM data transmission with 40 Gbit/s return-zero on-off keying (RZ OOK) signals. The result of this experiment was published in [C31] and is presented in Section 5.2.1. High-resolution spectroscopy revealed multiplet lines in each longitudinal comb mode, which impeded signal quality for data transmission. Nevertheless, it was the first WDM data transmission experiment using a Kerr comb source and an aggregate net data rate of 160 Gbit/s was demonstrated.

In Section 5.2.2 it is shown that a primary comb delivers high quality carriers for coherent data transmission as the multiplet problem is avoided. These experiments use a crystalline whispering gallery mode resonator from the group of Yanne Chembo at FEMTO-ST[14]. An aggregate net data rate of up to 400 Gbit/s [J1] is demonstrated. In Section 5.2.3 the technique of synchronizing sub-combs (Δ–δ-matching) [68] is applied to obtain low phase noise comb states. Using the second and third generation of SiN microresonators fabricated at EPFL, an aggregate net data rate of up to 1.35 Tbit/s is achieved and transmission over 300 km of standard single mode fiber is demonstrated [J4], [C15]. Finally, Section 5.2.4 discusses the results obtained using a soliton Kerr frequency comb. This comb allowed to transmit at 19.7 Tbit/s and bridging a distance of 75 km [C5].

[13] École Polytechnique Fédérale de Lausanne
[14] http://www.femto-st.fr/en/

5.2.1 On-off keying data transmission using a Kerr comb

The content of this Subsection is published in [C31].

Microresonator-Based Optical Frequency Combs for High-Bitrate WDM Data Transmission

Optical Fiber Communication Conference, *2012, paper OW1C.4.*

J. Pfeifle[1], C. Weimann[1], F. Bach[1], J. Riemensberger[3], K. Hartinger[3,5], D. Hillerkuss[1], M. Jordan[1], R. Holtzwarth[4,5], T. J. Kippenberg[3,4], J. Leuthold[1,2], W. Freude[1,2], C. Koos[1,2]

[1] *Institute of Photonics and Quantum Electronics (IPQ), Karlsruhe Institute of Technology (KIT), Germany*

[2] *Institute of Microstructure Technology (IMT), Karlsruhe Institute of Technology (KIT), Germany*

[3] *Swiss Federal Institute of Technology (EPFL) Lausanne*

[4] *Max Planck Institut für Quantenoptik (MPQ), Garching, Germany*

[5] *Menlo Systems GmbH, Martinsried, Germany*

A nonlinear high-Q SiN microresonator is used as a frequency comb generator for data transmission at 170.8 Gbit/s. The main sources for signal impairment are identified. Further dispersion engineering is crucial for Tbit/s transmission.

Introduction

Energy-efficient Tbit/s optical interconnects are key elements for future data communication among or within information processing systems. By using highly parallel wavelength division multiplexing (WDM) schemes with tens or hundreds of channels, high aggregate transmission capacity can be provided while keeping symbol rates compliant to the electrical bandwidth of CMOS driver circuitry [166]. Silicon photonics provides a platform for CMOS-compatible integration of the corresponding photonic-electronic interface [75]. However, while a Tbit/s chip-scale WDM receiver has been demonstrated recently [167], scalability at the transmitter side is still limited by the lack of adequate optical sources. Hybrid integration of III-V-dies on silicon is a promising approach to realize on-chip active devices [37], but dense integration of tens of narrowband DFB lasers is technologically challenging and subject to thermal constraints. Modulator-based schemes allow for flexible frequency comb generation [101] but need high-speed RF driving electronics and the number of carriers is inherently limited by the efficiency of the deployed phase shifters. Exploiting frequency conversion in a high-Q nonlinear microresonator, in contrast, allows to derive broadband comb spectra from a single continuous wave (CW) pump laser [66], [72]. A particularly interesting approach is the use of silicon nitride (SiN) waveguide technology [74], which allows for seamless integration of resonators with standard silicon photonic devices. Using frequency comb generation in a SiN ring, data transmission has recently been demonstrated for five individual carriers that have been selected from the comb spectrum and subsequently modulated with a 10 Gbit/s NRZ-OOK signal [168].

However, spectrally seamless data transmission on all lines of a frequency comb remains to be shown. One of the main difficulties is the strong variation in signal quality among the various optical carriers. In this paper we demonstrate modulation of the entire frequency comb with a 42.7 Gbit/s RZ-OOK signal, thereby transmitting an aggregate data rate of 170.8 Gbit/s on four channels. By systematically investigating the received data signals, we identify the occurrence of multiplet spectral lines as the main source of impairment. Interestingly, globally flat comb spectra are particularly prone to this effect, whereas combs with visually irregular spectra exhibit better signal quality. Careful dispersion engineering and advanced pumping schemes are crucial to avoid multiplet generation, thereby enabling Tbit/s data transmission.

Silicon nitride microresonator comb source

A microresonator with a 300 µm radius was used for parametric frequency comb generation, resulting in a free spectral range of 75 GHz. The waveguide has been designed to exhibit anomalous dispersion in the C-band for efficient four-wave mixing. It consists of a stoichiometric Si_3N_4 rib waveguide core which is embedded into an SiO_2 cladding. The core has a slightly trapezoidal cross section of 750 nm by 1700 nm with a sidewall inclination of 12° and is surrounded by a 50 nm thick Si_3N_4 slab, see Fig. 5.2(a). The loaded Q-factor is of the order of 500,000. Light is coupled to the chip via lensed fibers and inverse tapers with typical fiber-chip coupling losses of around 3 dB per facet.

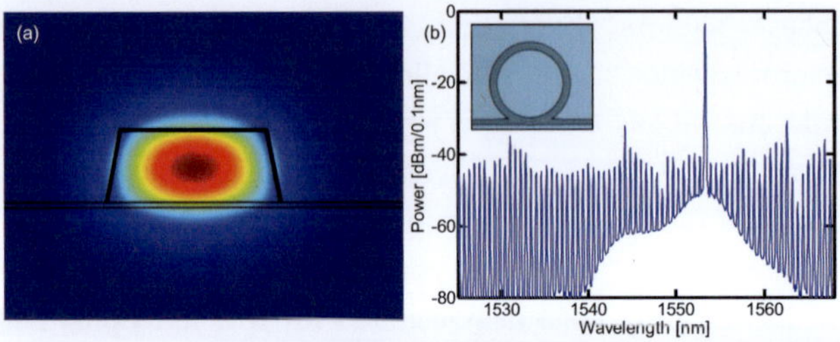

Fig. 5.2: (a) Simulation result of the TEM00 pump mode at 1552.4 nm. Strength of the radial electric field component is color-coded. (b) Typical obtained frequency comb spectrum, detail of C band. Inset shows microscope picture of silicon nitride ring resonator.

If a strong CW pump is coupled to one of the resonances, quantum fluctuations in neighboring resonances lead to parametric generation of new photons by Kerr nonlinearities. Multi-stage four-wave mixing (FWM) of the generated signal and idler waves and the pump then transfer energy to higher-order sidebands at the expense of the pump, hence generating a broadband frequency comb. A typical spectrum that is achieved with an on-chip pump power of approximately 30 dBm is shown in Fig. 5.2(b). The pump wavelength was tuned into a cavity resonance near 1553.3 nm. The four-wave mixing threshold decreases quadratically with increasing Q-factor [169]; it is therefore expected that the pump power can be significantly reduced in future devices with higher Q-factors and optimized coupling of the resonator to the waveguide.

WDM data transmission setup

The device was tested in a data transmission experiment. The setup is depicted in Fig. 5.3. A tunable CW laser and a high-power EDFA are used for pumping the ring resonator. The ASE noise of the pump EDFA is suppressed by a bandpass filter with a bandwidth of 2 nm. Comb lines with up to 0 dBm of power are achieved, while the comb is stable for several hours even without electronic feedback. In the transmitter, a wavelength-selective switch (WSS) is used to flatten the comb spectrum and suppress the pump. Two lithium-niobate Mach-Zehnder modulators are used in the transmitter to encode a 42.7 Gbit/s Return-to-Zero On-Off-Keying (RZ OOK) data signal with a 2^7-1 pseudo-random bit sequence (PRBS) onto the entire comb. At the receiver, individual channels are selected using a tunable bandpass filter (TBPF) with a bandwidth of 1.3 nm and a suppression of neighboring channels of more than 40 dB. A spectrum of a filtered 42.7 Gbit/s WDM channel is depicted in the inset in Fig. 5.3. A variable optical attenuator (VOA) is used to adjust the input power of the receiver's pre-amplifier. Received signals are analyzed with a digital communication analyzer (DCA) followed by a bit error ratio tester (BERT).

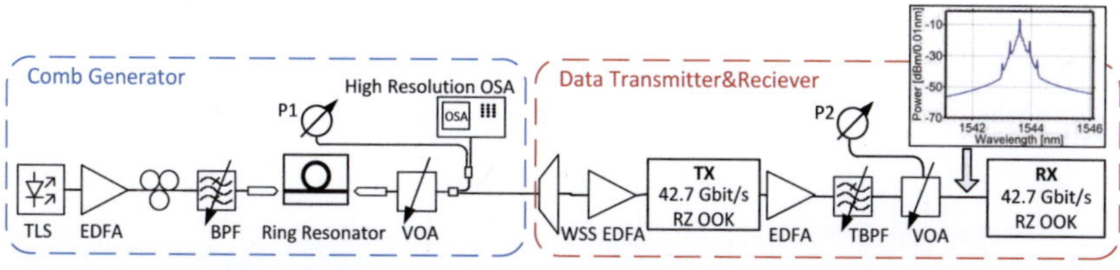

Fig. 5.3: Experimental setup for comb generation (blue) and data transmission (red). The inset spectrum shows a single channel extracted from the modulated comb. TLS: Tunable laser, EDFA: Erbium-doped fiber amplifier, (T)BPF: (tunable) bandpass filter, VOA: Variable optical attenuator, OSA: Optical spectrum analyzer, WSS: Wavelength-selective switch, TX: Transmitter, RX: Receiver with pre-amplification and BERT.

Analysis of carrier quality

Fig. 5.4(a) shows the spectrum of the comb used for data transmission. The resonance used for the pump is at 1553.3 nm. Interestingly, a complete, spectrally flat comb as depicted in Fig. 5.2 turned out to be unsuitable for data transmission. However, by carefully adjusting pump power and wavelength, parametric frequency conversion can be restricted to every third resonance line, resulting in a FSR of 1.8 nm rather than the native 0.6 nm. We have investigated the signal quality of various channels, five of which shall be discussed in more detail, see labels in Fig. 5.4(a). The dependence of the bit-error ratio (BER) on the receiver input power is depicted in Fig. 5.4(b); the corresponding eye diagrams are depicted in Fig. 5.4(c). The results are benchmarked to data transmission with a DFB laser ("DFB reference") coupled to the same transmitter/receiver setup. The quality of the received signals and the BER vary strongly between different comb lines. For carriers 1 and 2 no power penalty could be found within the measurement accuracy, and quality factors of $Q^2 = 17.7\,\text{dB}$ and $Q^2 = 18.0\,\text{dB}$ are measured. Carriers 3 and 4 exhibit an error floor at a BER of 10^{-4} and 10^{-6}, and quality factors are measured to be $Q^2 = 10.6\,\text{dB}$ and $Q^2 = 13.2\,\text{dB}$. Within the limits of forward error correction algorithms, the data transmission on these four channels can be considered error-free, summing up to an aggregate data rate of 170.8 Gbit/s transmitted on 4 carriers. Carrier 5, in contrast, does not exhibit an open eye, even though the optical power is comparable that of carrier 4.

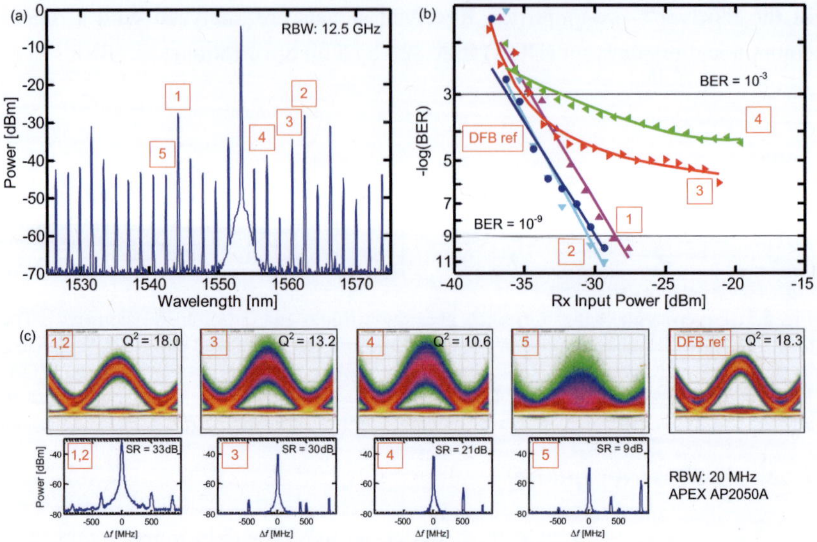

Fig. 5.4: (a) Spectrum of the frequency comb used for data transmission. The labels 1-5 correspond to carriers that were selected for exemplary investigations. Center wavelengths: 1: 1544.2 nm, 2: 1562.5 nm, 3: 1560.7 nm, 4: 1557.0 nm, and 5: 1542.3 nm. (b) BER vs receiver input power for four comb lines and reference DFB laser at 1550.9 nm for a 42.7 Gbit/s RZ OOK data signal. Channels 1 and 2 perform without power penalty, channels 3 and 4 exhibit an error floor but are still better than 10^{-3}. (c) Eye diagrams and Q-factors of the investigated channels together with high-resolution spectra of the carriers before modulation. SR denotes the suppression ratio of the main peak and the strongest side mode.

To identify the source of signal impairments, we have performed high-resolution spectral analysis of all carriers, see Fig. 5.4(c). We find that bad signal quality is caused by closely spaced multiplet spectral lines occurring within a single resonance, and this is also the reason which makes the visually neat comb in Fig. 5.2(b) unsuitable for data transmission. We believe that multiplet formation is a consequence of multi-stage four-wave mixing resulting from insufficient anomalous dispersion in the microresonator. Advanced pumping schemes and further engineering of the dispersion properties are therefore crucial to make resonator-based frequency combs viable for Tbit/s data transmission.

Summary

Simultaneous modulation of all comb lines of an integrated SiN microresonator frequency comb source with a 42.7 Gbit/s RZ OOK data signal and error-free reception of four channels was demonstrated, resulting in an aggregate data rate of 170.8 Gbit/s. The coexistence of multiple spectral lines within one resonance impedes transmission on other carriers. Advanced dispersion engineering and advanced pumping schemes are indispensable to suppress multiplet formation and make integrated frequency combs viable for Tbit/s data transmission.

5.2.2 Coherent communications using a primary comb

One possibility to prevent multiplet lines is to reduce the pump power such that only the primary comb lines appear. Such a primary comb corresponds to generating a Turing pattern inside the microresonator [170]. These patterns are very robust against fluctuations of the pump parameters and are highly coherent [171], [J1]. The journal publication [J1] represents a joint effort between KIT, group of Christian Koos, and FEMTO-ST, group of Yanne Chembo. It comprises two parts, a theoretical investigation that was conceived and carried out by Dr. Chembo and his group, and an experimental part, where a data transmission experiment was conducted at KIT using the WGM resonators fabricated at FEMTO-ST. This chapter contains the second part of [J1] describing the data transmission experiments.

Optimally Coherent Kerr Combs Generated with Crystalline Whispering Gallery Mode Resonators for Ultrahigh Capacity Fiber Communications

Phys. Rev. Lett. Vol. 114, Iss. 9, 093902 – Published 2 March 2015
DOI: 10.1103/PhysRevLett.114.093902

J. Pfeifle[1,2], A. Coillet[3], R. Henriet[3], K. Saleh[3], P. Schindler[1], C. Weimann[1], W. Freude[1,2], I. V. Balakireva[3], L. Larger[3], C. Koos[1,2], and Y. K. Chembo[3]

[1] *Institute of Photonics and Quantum Electronics (IPQ), Karlsruhe Institute of Technology (KIT), 76131 Karlsruhe, Germany*

[2] *Institute of Microstructure Technology (IMT), Karlsruhe Institute of Technology (KIT), 76131 Karlsruhe, Germany*

[3] *FEMTO-ST Institute (CNRS UMR6174), Optics Department, 15B Avenue des Monboucous, 25030 Besançon cedex, France*

We have performed a transmission experiment with primary combs originating from ultrahigh Q crystalline WGM resonators. The power spectrum used for our coherent transmission experiments, measured with a resolution bandwidth (RBW) of 20 MHz, is presented in Fig. 5.5. These combs correspond to the formation of a Turing pattern [172] in a microresonator. These patterns are very robust and strongly phase-locked [171]. In particular, their spectral purity is exceptionally high, and typically very close to that of the pumping laser. They can therefore be used as carriers in an ultrahigh capacity wavelength division multiplexing (WDM) network, as described below.

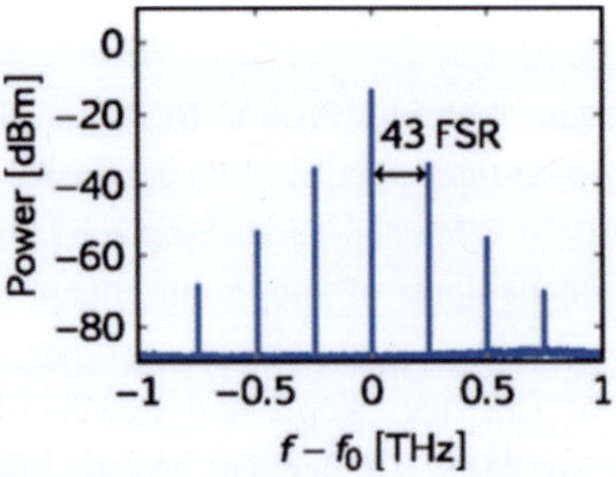

Fig. 5.5: Measured spectrum (RBW: 20 MHz) of the ultrastable Turing pattern used for the experiment. The resonator was a magnesium fluoride (MgF_2) disk with intrinsic quality factor $Q_{int} = 1.5\times10^9$ and group refractive index $n_{eg} = 1.37$ at the pump wavelength $\lambda_0 = 2\pi c_0/\omega_0 = 1552.0$ nm. Its diameter was $d = 11.3$ mm, yielding a free-spectral range $\Delta\omega/2\pi = 5.78$ GHz. In the figure, the Turing pattern of 43rd order has been excited, and the spectral distance between excited modes is therefore 248.54 GHz. The order of the Turing pattern can be tuned through the pump power or frequency [173]. The pump power was 295 mW, measured at the input of the tapered fiber which is coupled to the resonator with an extrinsic quality factor $Q_{ext} = 8.5\times10^9$.

The full experimental setup is displayed in Fig. 5.6. We use only three modes of the comb, namely the center mode (labeled as line 0) and its two immediately adjacent side modes (lines –1 and +1). We use quadrature phase shift keying (QPSK) or 16-state quadrature amplitude modulation (16QAM) as modulation formats, and combine them with Nyquist pulse shaping [44], [136]. At the receiver, digital signal processing is used for low-pass filtering, frequency offset compensation, clock recovery, polarization demultiplexing, dispersion compensation, and adaptive equalization. The quality of the received signal is evaluated by using the error-vector magnitude, which can be related to the bit error ratio [114].

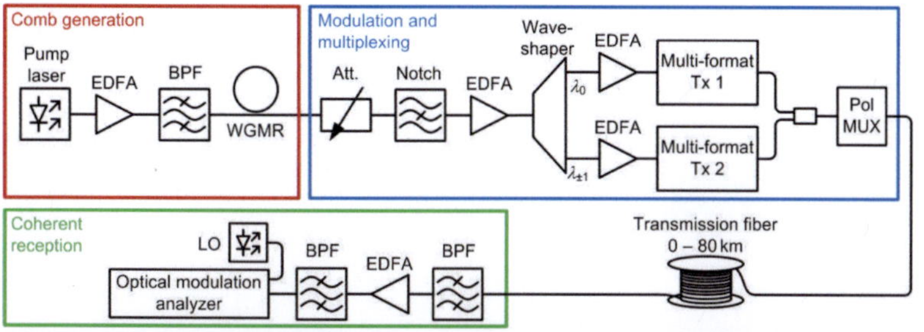

Fig. 5.6: Schematic representation of the experimental setup for data transmission. The Turing pattern generator comprises a tunable laser, an erbium-doped fiber amplifier (EDFA) and a bandpass filter (BPF) that rejects amplified spontaneous emission noise from the EDFA. A notch filter attenuates the center mode to a level comparable to the first-order side modes of the comb. A programmable filter is used for interleaving, flattening, and filtering the comb lines. Two EDFAs boost the carriers, which are fed via polarization controllers (PC) to two multiformat transmitters (Tx 1 for comb line 0, and Tx 2 for comb lines ± 1). After encoding of uncorrelated data onto the carriers at a symbol rate of 18 GBd, the two data streams are merged. To emulate polarization multiplexing (PolMUX), we create two time-delayed copies of the pseudorandom data stream and superimpose them on two orthogonal polarization states of the standard single-mode transmission fiber. We insert different lengths of transmission fiber, ranging from a few meters to 80 km. At the receiver, the data stream passes through bandpass filters and an EDFA before being coherently detected with an external cavity laser serving as a local oscillator (LO). The LO is tuned to match the center frequency of the respective channel and an optical modulation analyzer is used to monitor the quality of the received signal. The received data are visualized using constellation diagrams that display the real and the imaginary part of the optical amplitude in the complex plane.

The results of the data transmission experiments are summarized in Fig. 5.7. The 16QAM constellation diagrams for all wavelength channels and both polarizations are depicted along with the measured BER. Considering the symbol rate of 18 GBd on three wavelengths and two polarizations, we obtain an aggregate data rate of 432 Gbit/s. Using QPSK, we also demonstrate data transmission with up to 80 km of single-mode fiber without including additional amplification or any other means of signal regeneration. Fig. 5.7(b) shows the EVM as a function of distance for a transmission at 216 Gbit/s. For distances of up to 80 km, the EVM remains below 19.2%, which corresponds to a BER below 10^{-7} [114]. In both cases, error-free data transmission can be ensured if forward error correction with 7% overhead is considered.

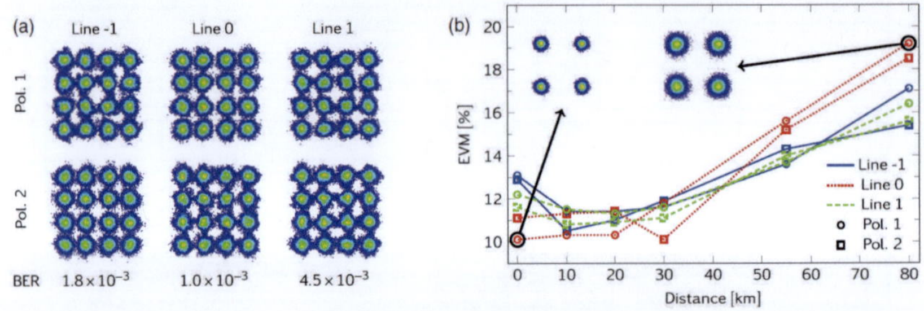

Fig. 5.7: Results of the data transmission experiments. (a) Back-to-back experiment at 144 Gbit/s per wavelength channel using 16QAM. The 16 constellation points are clearly distinguishable for each comb line and each polarization, indicating sufficiently low error-vector magnitude (EVM) and bit-error ratio (BER). After using forward-error correction (FEC) codes, a BER smaller than 4.5×10^{-3} guarantees error-free transmission $\left(\mathrm{BER}<10^{-15}\right)$. (b) Transmission experiment at 72 Gbit/s per wavelength channel using QPSK. The EVM is plotted as a function of the transmitted distance. The insets show constellation diagrams of the best and worst performance. For both 16QAM and QPSK, error-free data transmission can be ensured if FEC with 7% overhead is considered.

It is remarkable that data transmission on the side lines exhibits the same performance as on the center line, leading to data rates of 144 Gbit/s for each carrier—the highest value achieved for data transmission with Kerr combs so far. This unprecedented performance underlines the remarkable stability of the Turing patterns, which in this experiment did not require feedback stabilization. We may hence conclude that key performance parameters such as linewidth and relative intensity noise (RIN) are transferred from the pump to the newly generated modes. In this study, we have used only the first-order side modes (lines ±1), but higher-order spectral lines could be used as well. It is also important to note that optical Turing patterns are robust enough to enable transmission of data without further regeneration over 80 km.

In conclusion, we have demonstrated that optical Kerr frequency combs based on Turing patterns originating from a crystalline whispering-gallery-mode disk resonator are suitable sources for optical fiber data transmission, and allow for coherent data transmission using advanced modulation formats at up to 144 Gbit/s per wavelength channel. In particular, we anticipate that this comb generator could be of high interest for future optical fiber networks as it provides a path toward simple, compact, low-cost, energy-efficient, and reconfigurable multi-wavelength sources for optical WDM data transmission systems within or between data centers [52].

5.2.3 Coherent communications enabled by Δ–δ-matching

While the primary combs are suitable to carry 16QAM data signals, they have the drawback that only very few comb lines are generated and that the power per line decreases exponentially towards the red and the blue side of the pump. Furthermore, the line spacing corresponds to an integer multiple of the cavity FSR. In order to obtain a primary comb with a line spacing that allows to generate a WDM data stream with high spectral efficiency, it would be required to either decrease the cavity FSR or to strongly increase its dispersion (see Appendix A.2).

In order to increase the usable number of lines, it is preferable to generate a comb line in each resonance while avoiding the formation of multiplet lines. This has been demonstrated for the first time in [68] by fine-tuning of the pump parameters in order to achieve $\Delta - m \cdot \delta = 0$, see Fig. 5.1. In the following we experimentally demonstrate that these combs are suitable for coherent data transmission with the modulation formats QPSK and 16QAM. We demonstrate a scheme to stabilize the comb states for extended data transmission experiments. Additionally we present a signal impairment analysis and discuss options for further improvement of the comb. The contents of this section are published in [J4] and its Supplementary Information. Furthermore the results were presented at three conferences [C14], [C15], and [C23].

Coherent terabit communications with microresonator Kerr frequency combs

Nature Photonics, Volume 8, Issue 5, pp. 375–380, 2014
DOI: 10.1038/nphoton.2014.57

J. Pfeifle,[1] V. Brasch,[2] M. Lauermann,[1] Y. Yu,[1] D. Wegner,[1] T. Herr,[2] K. Hartinger,[3] P. Schindler,[1] J. Li,[1] D. Hillerkuss,[1,†] R. Schmogrow,[1] C. Weimann,[1] R. Holzwarth,[3] W. Freude,[1] J. Leuthold,[1,†] T. J. Kippenberg,[2] C. Koos[1]

[1] *Institute of Photonics and Quantum Electronics (IPQ) and Institute of Microstructure Technology (IMT), Karlsruhe Institute of Technology (KIT), 76131 Karlsruhe, Germany*

[2] *Ecole Polytechnique Federale de Lausanne (EPFL), 1015 Lausanne, Switzerland*

[3] *Menlo Systems GmbH, 82152 Martinsried, Germany*

[†] *now with Electromagnetic Fields & Microwave Electronics Laboratory (IFH), ETH Zurich, 8092 Zurich, Switzerland*

Optical frequency combs have the potential to revolutionize Tbit/s communications [50]. Generation of Kerr combs in nonlinear microresonators [66] represents a particularly promising option [168] enabling line spacings of tens of GHz. However, such combs may exhibit strong phase noise [68], [174], [C31], which has made high-speed data transmission impossible up to now. Here we demonstrate that systematic adjustment of pump conditions for low phase noise [68], [71], [160], [175] enables coherent data transmission with advanced modulation formats that pose stringent requirements on the spectral purity of the comb. In a first experiment, we encode a data stream of 392 Gbit/s on a Kerr comb using quadrature phase shift keying (QPSK) and 16-state quadrature amplitude modulation (16QAM). A second experiment demonstrates feedback-stabilization of the comb and transmission of a 1.44 Tbit/s data stream over up to 300 km. The results show that Kerr combs meet the highly demanding requirements of coherent communications and thus offer an attractive solution towards chip-scale Tbit/s transceivers.

Experimental setup and resonator design

The pump for Kerr comb generation is generated by an external-cavity laser (New Focus Velocity Model TLB-6728), a polarization controller (PC) and an erbium-doped fiber amplifier (EDFA) providing an output power of up to 37 dBm, see Fig. 5.8. The coupling loss between the lensed fiber and the Si_3N_4

chip amounts to approximately 3 dB per facet. We pump a resonance near 1549.4 nm using an on-chip pump power of approximately 33 dBm. The tunable fiber Bragg grating (FBG 1) at the output of the device suppresses the remaining CW pump wave by approximately 20 dB. The ring resonator consists of nearly stoichiometric Si_3N_4 grown in multiple layers with intermediate annealing steps [74]. The strip waveguides are patterned with electron-beam lithography and transferred to the substrate by reactive ion etching with SF_6/CH_4 chemistry. After etching, the structures are embedded into a SiO_2 cladding using low-pressure chemical vapor deposition (LPCVD). The waveguides are 2 µm wide and 750 nm high, and the sidewalls are inclined by 12° with respect to the vertical direction, see Fig. 5.8, Inset 1 which shows the waveguide cross section of the resonator and the calculated mode profile.

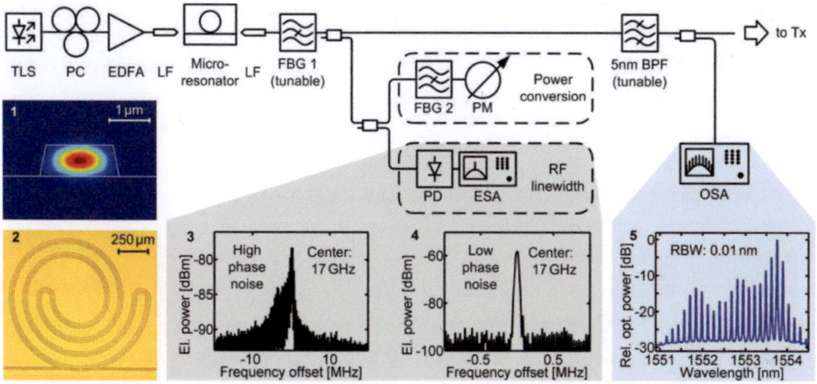

Fig. 5.8: Comb generation setup. The optical pump comprises a tunable laser source (TLS), a polarization controller (PC), and an erbium-doped fiber amplifier (EDFA). Lensed fibers (LF) couple light to and from the microresonator chip. A fiber Bragg grating (FBG 1) serves as a tunable narrowband notch filter to suppress residual pump light. For adjustment of the pump parameters, we monitor the power conversion from the pump to the adjacent lines (PM: Power meter). An electronic spectrum analyzer (ESA) is used to measure the RF linewidth in the photocurrent spectrum of the photodetector (PD). A 5 nm-wide spectral section is extracted from the comb spectrum and used for data transmission. Insets: **1,** waveguide cross section and mode profile; **2,** optical micrograph of the resonator; **3,** RF spectrum of a high phase-noise comb state (RBW 10 kHz); **4,** RF spectrum of a low phase-noise comb state (RBW 30 kHz); **5,** Selected part of the comb spectrum (OSA: Optical spectrum analyzer).

The high-temperature growth technique used to fabricate the 750 nm thick layers of stoichiometric Si_3N_4 requires annealing at 1200 °C. This makes it difficult to fabricate SiN resonators in the framework of standard CMOS processes. One option to overcome these limitations is to use dedicated fabrication processes and multi-chip integration as illustrated in Fig. 1.6, Section 1.2.3. Alternatively, it is possible to deposit Si_3N_4 at 400 °C and to use UV thermal processing (UVTP) at lower temperatures to reduce the defect density. Both techniques are subject to ongoing research.

The ring resonator waveguide is coiled up to reduce the footprint and exhibits a loaded Q-factor of 8×10^5. In contrast to the concept illustrated in Fig. 1.6, our experiments rely on comb generators with a single waveguide to couple pump light to the resonator and to extract the frequency comb. As a consequence, strong CW pump light can pass through the resonator and needs to be suppressed by a tunable fiber Bragg grating (FBG 1) at the output of the device. It has recently been shown that stability and phase noise of the Kerr comb are closely linked to the pump conditions [68]. Careful tuning of pump power, frequency, and polarization is therefore of prime importance. To adjust these parameters in the experiment we use two optimization criteria: The power of the newly generated comb lines measured behind a second fiber Bragg grating (FBG 2), and the radio frequency (RF) linewidth measured using a photodetector (PD) and electrical spectrum analyzer (ESA).

Pumping scheme for low phase-noise Kerr combs

To obtain low phase-noise Kerr combs, the pump parameters are adjusted in two steps using the setup depicted in Fig. 5.8: First, the pump wavelength is periodically scanned across the resonance at a frequency of approximately 100 Hz while staying within the stop band of FBG 2 and continuously measuring the power conversion to the spectral region outside the stop band. During these scans the polarization of the pump signal is slowly varied to maximize the conversion. Maintaining this polarization, the detuning of the pump signal with respect to the resonance wavelength is then carefully adjusted until the initially broad RF spectrum (Fig. 5.8, Inset 3) exhibits a single narrow peak (Fig. 5.8, Inset 4). Note that the frequency axes of Insets 3 and 4 have different scales. For an on-chip pump power of approximately 33 dBm, the essential part of the optical comb spectrum is depicted in Inset 5 of Fig. 5.8. The comb lines for the data transmission experiment were selected from this partial spectrum.

Experimental setup for WDM data transmission

To emulate a wavelength division multiplexed (WDM) communication system, we transmit data on all channels simultaneously, using one pseudo-random bit sequence (PRBS) on the even channels and another independent PRBS on the odd channels. Neighboring channels hence carry uncorrelated data streams, and inter-channel crosstalk translates directly into a degradation of signal quality. The corresponding setup is depicted in Fig. 5.9. Carriers belonging to odd and even channels are first separated by a tunable delay interferometer (DI) [176]. The superposition of both spectra is displayed in Inset 1 of Fig. 5.9. Two home-built multi-format transmitters [136] are then used to encode QPSK or 16QAM signals with a symbol rate of 14 GBd on the sets of even and odd carriers using independent PRBS with a length of $2^{15}-1$. Polarization-division multiplexing (PDM) is emulated by splitting the optical signal, and recombining the original and a sufficiently delayed copy to form two decorrelated data streams on orthogonal polarizations of a standard single-mode fiber (SSMF). At the receiver, the signal is amplified, filtered and coherently detected using an optical modulation analyzer (OMA, Agilent N 4391 A) with a standard tunable laser (Agilent 81680A) acting as a local oscillator (LO). An optical bandpass filter with a 2 nm passband is used to pre-select the spectral region of interest and to avoid saturation of the receiver photodiodes. Individual WDM channels are addressed by appropriate choice of the LO wavelength and by digital brick-wall filtering in the baseband. Inset 2 of Fig. 5.9 shows the optical spectrum at the input of the OMA for the first data transmission experiment. The effect of the 2 nm band-pass filter is clearly visible; the noise in the pass-band predominantly originates from the ASE of the pump EDFA that is used for frequency comb generation.

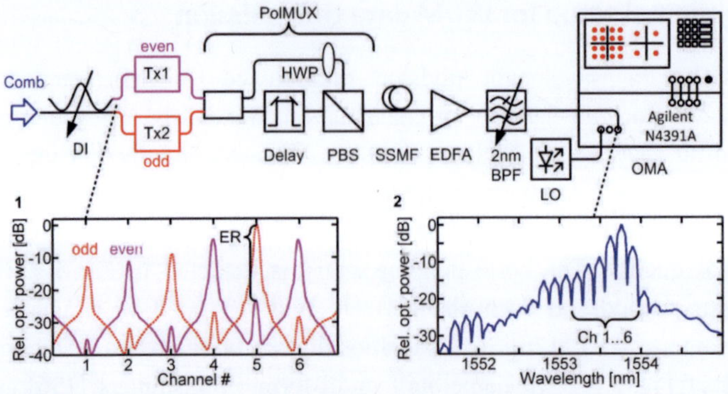

Fig. 5.9: Experimental setup for the first data transmission experiment. A tunable delay interferometer (DI) separates the frequency comb into odd and even carriers. The two sets of carriers are then independently modulated at a symbol rate of 14 GBd using QPSK and 16QAM signals and Nyquist sinc-pulses (Tx: Transmitter). The two data streams are merged and fed into a polarization multiplexer (PolMUX), comprising a half-wave plate (HWP), a delay line, and a polarization beam splitter (PBS). The signal is sent to the receiver using standard single-mode fiber (SSMF). The receiver comprises an erbium-doped fiber amplifier (EDFA), an optical band-pass filter (BPF), and an optical modulation analyzer (OMA) with a tunable laser serving as local oscillator (LO). Insets: **1**, Spectra of the unmodulated carriers after separation by the DI. The extinction ratio (ER) is better than 20 dB for all channels; **2**, Optical spectrum at the input of the OMA, after the 2 nm band-pass filter.

Digital post-processing comprises demultiplexing of the polarization channels, compensation of the frequency offset between the LO and the optical carrier, clock recovery, and equalization. For performance evaluation we use the error-vector magnitude (EVM) metric, which describes the effective distance of a received complex symbol from its ideal position in the constellation diagram. Provided that the signal is impaired by additive white Gaussian noise only, the EVM is directly related [114], [115] to the bit-error ratio (BER). The BER threshold for second-generation forward error correction [138] (FEC) is given by 4.5×10^{-3}.

QPSK and 16QAM at 14 GBd using a 17 GHz Kerr comb

The results of the data transmission experiment are summarized in Fig. 5.10. Fig. 5.10(a) depicts the optical power spectrum of the modulated carriers for all six data channels. We did not flatten the comb spectrum prior to modulation for comparing the influence of different carrier powers on the transmission performance. When using a standard forward error-correction (FEC) scheme with 7 % overhead, the limits for error-free detection are given by an EVM of

38 % (11 %) for QPSK (16QAM), see Appendix B.4.3. The constellation diagrams for each polarization of the six wavelength channels are depicted in Fig. 5.10(b) along with the measured EVM. All channels are well below the 38 % threshold for QPSK and channel 5 even shows well enough performance for 16QAM transmission. We transmit with a symbol rate of 14 GBd and choose QPSK for channels 1…4 and 6, and 16QAM for channel 5. Taking into account polarization multiplexing, we obtain an aggregate data rate of 392 Gbit/s. Considering the overhead of 7 % for FEC, the net spectral efficiency amounts to 3 bit/s/Hz for the QPSK and to 6 bit/s/Hz for the 16QAM channels.

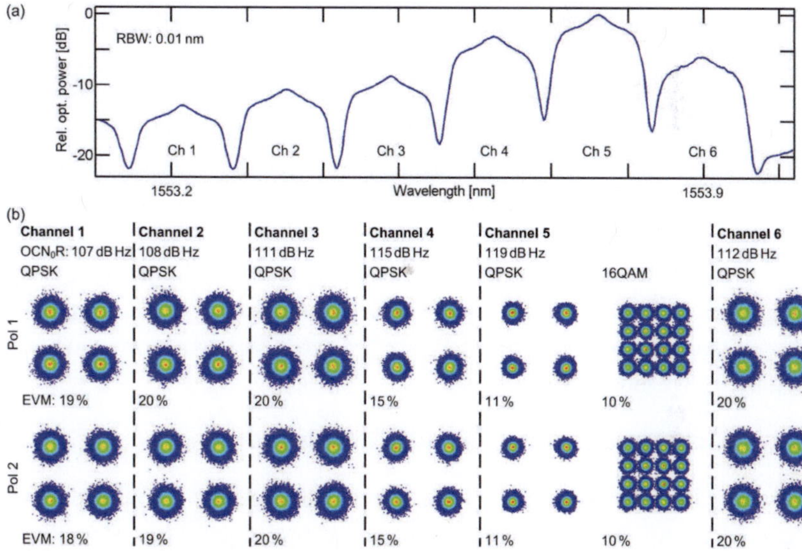

Fig. 5.10: Coherent data transmission using a Kerr microresonator frequency comb. (a) Spectrum of modulated carriers for all six data channels, measured at the input of the optical modulation analyzer (OMA). (b) Constellation diagrams for each channel and for both polarizations along with the corresponding error vector magnitude (EVM). The constellation diagrams show no sign of excessive phase noise, which would result in constellation points that are elongated along the azimuthal direction. For QPSK, the BER of all channels is below 4.5×10^{-3}, which corresponds to an EVM of 38 %; for channels 4 and 5 the BER is even smaller than 10^{-9} (EVM < 16.7 %). The good quality of channel 5 enables transmission of a 16QAM signal with a measured BER of 7.5×10^{-4}.

When relating the EVM to the bit error ratio (BER), we assume that the signal is impaired by additive white Gaussian noise. The validity of this assumption is supported by the fact that the deviations of the measured from the ideal constellation points occur equally in all directions and do not show any sign of anisotropy. For 16QAM, we estimate a BER of 3×10^{-4} from the measured EVM of 10%, which is in fair agreement with the measured BER of 7.5×10^{-4}. This underlines the validity of the EVM-based estimation of the BER in our experiment. For the other QPSK channels, the BER of 3×10^{-7}, 7×10^{-8}, 1×10^{-8}, 1×10^{-11}, and 5×10^{-20} are estimated accordingly from the measured EVM values of 20 %, 19 %, 18 %, 15 %, and 11 %. Channels 4 and 5 hence feature BER that are well below the threshold of 10^{-9}, which is commonly considered error-free in telecommunications. This clearly demonstrates that error-free data transmission with Kerr combs is possible and that the achievable BER is limited by the components of the transmission system such as modulators, amplifiers and filters rather than by the coherence of the light source.

Excessive phase noise, in particular, can be excluded as a relevant impairment of data transmission as this would lead to constellation points which are elongated along the azimuthal direction. Instead, the signal impairments can be attributed to strong amplified spontaneous emission (ASE) originating from the high power pump EDFA. In the current configuration, ASE light passes straight through the resonator chip and superimposes the comparatively weak comb lines as additive white Gaussian noise. The results indicate that Kerr combs are indeed perfectly suited for data transmission with phase-sensitive modulation formats.

Tbit/s data transmission using a 25 GHz Kerr comb

In a second experiment, we use a resonator with a FSR of 25 GHz and a Q-factor of 2×10^{6}, which is higher than in the first experiment. This allows reducing the pump power to a level where filtering of the ASE noise from the EDFA is possible. To enable stable long-term operation of the Kerr comb for extended data transmission experiments, we implement a feedback loop, which locks the wavelength of the pump laser to a specified position within the resonance of the cavity, as described in the subsequent section. In contrast to the first WDM experiment, we now flatten the comb lines prior to modulation using a programmable filter. For data transmission we use QPSK at a symbol rate of 18 GBd, and we insert up to four fiber spans of 75 km length between the transmitter and the receiver. The results are summarized in Fig. 5.11. We obtain 20 channels with an EVM below the threshold of 38 %. The total data stream amounts to 1.44 Tbit/s with a net spectral efficiency of 2.7 bit/s/Hz.

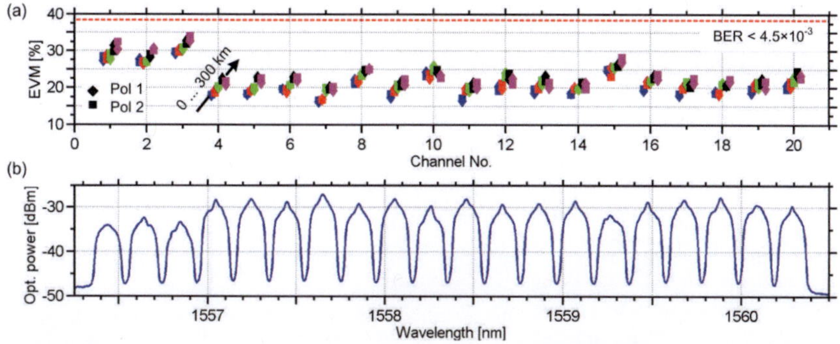

Fig. 5.11: Coherent Tbit/s data transmission using a feedback-stabilized Kerr frequency comb. (a) EVM for all data channels and fiber spans. The carriers are modulated at a symbol rate of 18 GBd using QPSK and Nyquist pulse shaping. Using polarization multiplexing at each of the 20 WDM channels, an aggregate data rate of 1.44 Tbit/s is achieved. The polarizations are distinguished by diamonds and squares, while the different fiber spans (0 km, 75 km, 150 km, 225 km, 300 km) are color-coded and slightly offset in the horizontal direction as indicated by the arrow. The red dashed line indicates an EVM of 38 %, which corresponds to the BER threshold for second-generation FEC. (b) Optical spectrum of the 1.44 Tbit/s data stream. The spectrum was flattened prior to modulation.

Feedback stabilization

To maintain low phase-noise comb states during the second experiment, we implemented a control loop that uses the power P_{comb} measured at the output of FBG 2 as a feedback signal and adjusts the wavelength of the pump laser to ensure a constant detuning. Fig. 5.12(a) shows the comb generation setup along with the control loop. The optical pump is generated and adjusted using the power of the newly generated comb lines and the radio-frequency (RF) linewidth of the baseband beat signal as optimization criteria. The resonator exhibits a free spectral range (FSR) of 25 GHz and a loaded Q-factor of 2×10^6 – about a factor of 2.5 larger than the Q-factor of the resonator used in the first experiment. This allows to reduce the pump power and to add a 2 nm band-pass filter (BPF) before the resonator chip to suppress ASE noise from the pump amplifier. The on-chip power amounts to 29 dBm and is coupled to a resonance near 1549.4 nm.

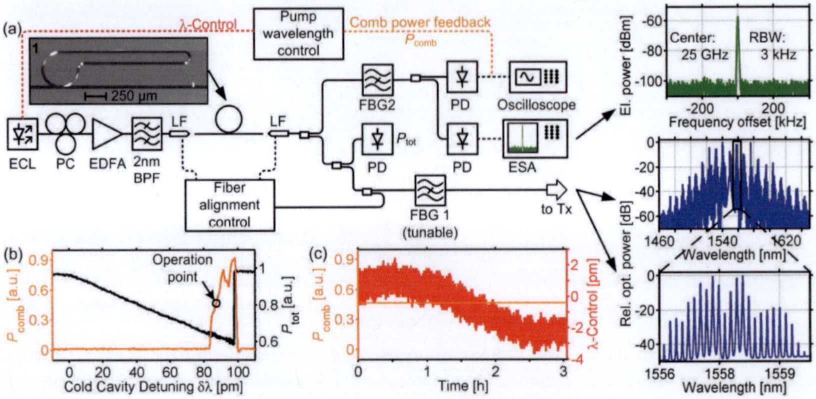

Fig. 5.12: Generation and stabilization of the Kerr frequency comb in the second experiment. (a) The setup is similar to the one used in the first experiment (ECL: External cavity laser, PC: Polarization controller, EDFA: Erbium-doped fiber amplifier, BPF: Band-pass filter, LF: Lensed fiber, FBG: Fiber Bragg grating, PD: Photodetector, ESA: Electrical spectrum analyzer, Tx: Transmitter). An improved Q-factor allows reducing the pump power and filtering the ASE noise from the EDFA. The insets show the RF (top) and optical (mid and bottom) spectra of the resulting comb. For stable comb output, the power P_{comb} behind FBG2 is used as feedback for controlling the pump wavelength. (b) P_{comb} (left axis, orange) and total power P_{tot} measured at the chip output (right axis, black) vs. detuning $\delta\lambda$. (c) Feedback and wavelength control signal recorded over three hours. The pump wavelength follows thermal drifts of the resonance, leading to perfectly stable output power.

The principle of the feedback stabilization is illustrated in Fig. 5.12(b), which shows P_{comb} (orange trace, left axis) and the total chip output power P_{tot} (black trace, right axis) as a function of the detuning $\delta\lambda$ between the pump wavelength and the resonance wavelength of the cold cavity. As the wavelength of the pump is scanned, it drags along the resonance through thermal locking. This leads to a continuous decrease of P_{tot} until locking is lost for large detuning. At the same time, P_{comb} exhibits a pronounced increase once the power coupled to the cavity reaches the threshold for Kerr comb generation. Our control loop operates at the rising edge of P_{comb} and is based on a commercially available PID controller (TEM Messtechnik LaseLock). The circle in Fig. 5.12(b) indicates the operating point that corresponds to the low phase-noise comb state for which the pump parameters have been manually optimized before starting the control. Moreover, for stable comb generation, it is crucial that also the fiber-chip coupling efficiency is actively stabilized. To this end, we use an independent control loop based on commercially available hardware (Melles Griot NanoTrak) to keep the lensed fibers in the optimum coupling position. We confirm the functionality of the feedback stabilization by recording the feedback signal and the comb power for more than 3 h, Fig. 5.12(c). The pump wavelength follows the thermal drifts of the resonance, and residual comb power fluctuations are found to be well below 1 %. Comb spectra measured at the beginning and at the end of the extended data transmission experiment are indistinguishable within the accuracy of our spectrometer (ANDO AQ 6317B).

Signal impairment analysis

In the first experiment, ASE from the pump EDFA passes straight through the resonator chip and superimposes the comparatively weak comb lines as additive white Gaussian noise. The power of the comb lines is limited by the efficiency of the frequency conversion and the coupling of the pump to the resonator. As a figure of merit we define the optical carrier-to-noise-density power ratio (OCN_0R), which relates the power of the unmodulated carrier to the underlying noise power in a bandwidth of 1 Hz. To determine the OCN_0R, we need to estimate the noise power density right at the frequency of the respective comb line. This is done by averaging the spectral power densities obtained in the middle between two comb lines. The spectral power densities are measured with an optical spectrum analyzer (Ando AQ 6317B) in a 0.01 nm bandwidth and renormalized to a bandwidth of 1 Hz. For channel 5, we obtain a good OCN_0R of more than 119 dB Hz, which enables a very good QPSK signal and a good 16QAM signal. Channel 4 with an OCN_0R of more than 115 dB Hz still gives a very good QPSK signal but lacks performance for 16QAM. Channels 1…3 as well as channel 6 exhibit even lower OCN_0R resulting in larger EVM figures. Because of this correlation between the OCN_0R and the EVM we conclude that additive white Gaussian noise is the major source of signal impairment in the first data transmission experiment.

To improve the signal quality, it is possible to filter out ASE by using a band-pass filter before the resonator chip as demonstrated in the second experiment. This was not possible in the first experiment, since we did not have any filters that could safely handle the 37 dBm of pump power right after the EDFA. For the second experiment, we have further optimized the fabrication process, thereby increasing the microresonator's Q-factor and hence decreasing the required threshold power [169], such that filtering of the pump signal becomes possible. In future experiments, the comb may be extracted by a second waveguide which is coupled to the microresonator [177]. This would avoid direct transmission of broadband ASE noise through the resonator device. Using resonators with higher Q-factor might even enable to use a high-power pump laser diode without requiring external amplification at all.

Apart from the ASE of the pump EDFA, we investigated coherent crosstalk as a potential source of signal degradation: In the first experiment, the tunable DI used to separate the carriers has a finite extinction ratio (ER), see Inset 1 in Fig. 5.9. As a consequence, the data signals of the even channels are superimposed by residual carriers modulated with odd-channel data and vice versa, which leads to further signal degradation. However, in our experiment this effect was not relevant: With an ER better than (22.2 ± 1.5) dB for all channels, no correlation of channel ER and EVM can be observed. In the second experiment we use a programmable filter, thereby achieving an even higher ER of (24.6 ± 4.0) dB. Coherent crosstalk can therefore safely be excluded as a source of signal degradation in both experiments.

5.2.4 Coherent communications using a soliton comb state

Although the method of Δ–δ-matching can solve the multiplet problem and increase the number of useful comb lines as compared to using a primary comb, it still suffers from large variations of the power per comb line. In other words, the spectral flatness of these combs is extremely poor. First demonstrated in 2014 [71], it is also possible to generate solitons inside the micro-cavity. These solitons exhibit a sech^2-shaped spectral envelope and thus provide a large number of comb lines with little variations of the power per comb line. Details on the excitation of solitons in microresonators are discussed in Appendix A.3. Details of the design and the fabrication process for the microresonator used for the following experiment are published in [70] and [178], respectively. The content of the following subsection is published in [C5] and comprises the first data transmission experiment using a soliton comb state. Furthermore, it is the first experiment using a microresonator-based comb, where the achieved data rate is not limited by the comb source but by the available transmission equipment in our laboratories.

Full C and L-Band Transmission at 20 Tbit/s Using Cavity-Soliton Kerr Frequency Combs

Conference on Lasers and Electro-Optics, *2015, postdeadline paper JTh5C8.*

J. Pfeifle[1], A. Kordts[2], P. Marin[1], M. Karpov[2], M. Pfeiffer[2], V. Brasch[2], R. Rosenberger[1,3], J. Kemal[1], S. Wolf[1], W. Freude[1,3], T. J. Kippenberg[2], C. Koos[1,3]

[1] *Institute of Photonics and Quantum Electronics (IPQ), Karlsruhe Institute of Technology (KIT), Germany*

[2] *École Polytechnique Fédérale de Lausanne (EPFL), Lausanne, Switzerland*

[3] *Institute of Microstructure Technology (IMT), Karlsruhe Institute of Technology (KIT), Germany*

Temporal soliton formation in a Kerr-nonlinear optical microresonator leads to a coherent frequency comb covering the C and L-band. We demonstrate WDM transmission with 94 comb lines and an aggregate data rate of 20 Tbit/s.

Introduction

Massively parallel wavelength-division multiplexing (WDM) links will become key elements of future metro and data-center networks. As an alternative to arrays of individually stabilized lasers, frequency combs are currently investigated as optical sources for such links, providing a large number of carriers which can be stabilized by fixing the comb's center frequency and the line spacing. So far, frequency comb generation for ultra-fast data transmission relied either on conventional solid-state mode-locked lasers combined with spectral broadening in highly nonlinear fibers (HNLF) [52], or on multi-stage parametric mixers seeded with the modulated output of an external-cavity laser [53]. However, these comb generation schemes require complex setups with discrete devices, and the potential for chip-scale miniaturization is limited. This can be overcome by broadband frequency combs generated in Kerr-nonlinear microresonators [72]. Kerr frequency combs exhibit low phase and amplitude noise, and were already used for coherent transmission [J4], but strong spectral variations of the comb limited the data rate to 1.44 Tbit/s.

In this paper we use a novel regime of Kerr comb generation that exploits single dissipative solitons [71], in a microresonator pumped with a continuous-wave (CW) laser. Fully coherent sech^2-shaped comb spectra with hundreds of lines span multiple communication bands with an unprecedented flatness. In our experiment, a total of 94 lines in the C and L-band were used to transmit data streams of 224 Gbit/s (28 GBd PDM-16QAM) per line, leading to a line rate (net data rate) of 21.1 Tbit/s (19.7 Tbit/s). This is the first time that a cavity-soliton Kerr comb was used for data transmission, achieving by far the highest data rate for a chip-scale frequency comb source.

Comb generation and characterization

Parametric frequency conversion in Kerr-nonlinear microresonators generates broadband combs from single CW pumps [72]. Recently, the formation of temporal cavity solitons [71] has drawn much attention. Temporal solitons exist when a resonator with second-order anomalous group velocity dispersion is pumped far above threshold, and if mode crossings near the pump frequency are avoided [69]. Fig. 5.13(a) shows the setup for soliton generation. Pump light from an external cavity laser (ECL) is frequency-tuned via an analog signal generated by an arbitrary waveform generator (AWG). We adjust polarization and power of the pump (~3 W) using a polarization controller (PC) and an erbium-doped fiber amplifier (EDFA), respectively. Amplified spontaneous emission (ASE) noise is suppressed by a 0.8 nm wide band pass filter (BPF). Lensed fibers (LF) couple light to and from the chip with insertion losses of 1.5 dB per facet at the pump wavelength of 1548.76 nm. After the chip, a fiber Bragg grating (FBG) and an optical isolator suppress residual pump light. The pump parameters are tuned as described in [71]. The pump power is monitored with a photodiode (PD) and an oscilloscope. Fig. 5.13(b) shows an SEM picture of the ring resonator having a free spectral range of 95.8 GHz and a Q-factor of $\sim 10^6$. Fig. 5.13(c) shows the spectrum of the generated comb after the FBG. Within the C and L-band, the comb lines have a power of (-18.7 ± 5.4) dBm. The average optical carrier-to-noise power ratio (OCNR) of > 57.0 dB (noise in 0.1 nm, optical carrier-to-noise-density power ratio $OCN_0R > 158$ dB Hz), is limited by the noise floor of our optical spectrum analyzer (OSA). The inset of Fig. 5.13(c) shows the comb lines near the pump. Their OCNR is degraded due to the ASE noise passing the 0.8 nm BPF. This renders the two nearest neighbors of the pump useless for data transmission in the current experiment.

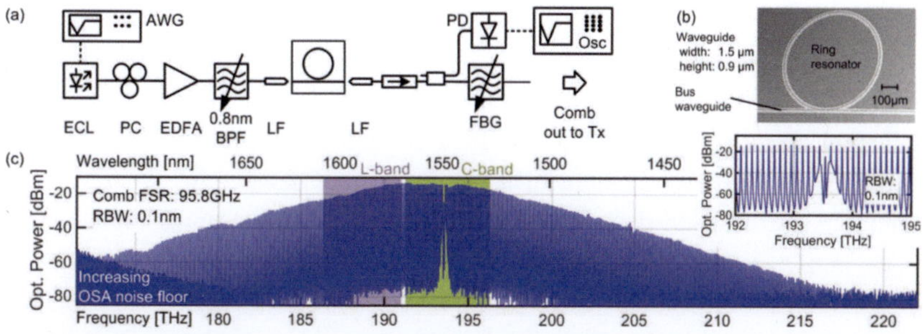

Fig. 5.13: Cavity-soliton Kerr comb generation in a SiN microresonator. (a) Experimental setup. (b) SEM picture of the ring resonator. (c) Comb spectrum after FBG (RBW 0.1 nm). The inset shows the comb lines near the suppressed pump.

20 Tbit/s transmission experiment using C and L-band

To prove the viability of a cavity-soliton Kerr comb for high-speed data transmission, we emulate a wavelength-division multiplexing (WDM) transmission system as depicted in Fig. 5.14. The comb is amplified using a C+L-band EDFA and split in two paths. Two Finisar WaveShapers (WS) cover the L-band (1569.6 nm to 1608.7 nm) and the C-band (1527.4 nm to 1567.5 nm), respectively, and separate the comb into even and odd carriers. Additionally, the WS flatten the comb and compensate the gain tilt of the subsequent EDFA. A first 3dB-coupler combines the odd lines in the C and L-band, and a second one combines the even lines of both bands. The two sets of carriers are independently modulated at 28 GBd using 16-state quadrature amplitude modulation (16QAM) with a pseudo-random bit sequence of length $2^{11}-1$. Polarization division multiplexing (PDM) is emulated as in [J4].

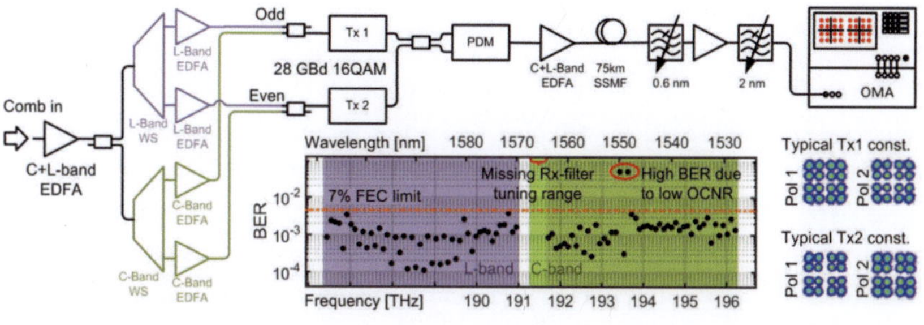

Fig. 5.14: Data transmission setup, measured BER and representative constellation diagrams.

In the transmission experiment, we amplify the data stream to a power of 17 dBm per spectral band using a C+L-band EDFA and send it over 75 km of SSMF. An optical modulation analyzer (OMA, Keysight N4391A) serves as coherent receiver with an internal ECL as a local oscillator (LO). During offline processing, we perform filtering, frequency offset compensation, clock recovery, polarization demultiplexing, dispersion compensation, and adaptive equalization, before we determine the bit-error-ratio (BER) for each channel. The BER for all channels and a representative set of constellation diagrams are shown in the lower right of Fig. 5.14. All but two channels had a BER below 4.5×10^{-3}, the limit for forward error correction (FEC) with an overhead of 7% [179]. This threshold is only exceeded by the two channels at 193.4 THz and 193.6 THz, which are nearest neighbors of the pump and hence feature too low an OCNR, see Fig. 5.13(c). At the low-frequency end of the C-band, five channels could not be characterized due to limited tuning range of the Rx filter. The other boundaries are given by the spectral window of the WS. Counting only the 94 measured channels, we obtain an aggregate line rate (net data rate) of 21.1 Tbit/s (19.7 Tbit/s).

Summary

We demonstrate for the first time that an optical frequency comb generated by a single temporal soliton in a SiN microresonator can be used for massively parallel WDM transmission. In a proof-of-principle experiment we use 94 lines in the C and L-band to transmit 20 Tbit/s over a distance of 75 km. We expect that data rates can be boosted towards 100 Tbit/s by decreasing line spacing, increasing symbol rate, and by using additional transmission bands.

5.3 Summary and conclusion

In summary, Kerr frequency combs are shown to be well suited for high-capacity data transmission with phase-sensitive modulation formats. Error-free transmission with data rates of up to 19.7 Tbit/s, spectral efficiencies of up to 6 bit/s/Hz, and transmission distances of up to 300 km are demonstrated. The received signals exhibit no sign of excessive phase noise. Assuming that the demonstrated spectral efficiency of 6 bit/s/Hz can be maintained over the entire bandwidth of the Kerr comb, chip-scale transmitters providing aggregate data rates beyond 100 Tbit/s are envisioned, only limited by nonlinear effects in the single-mode silica fibers used for transmission [7]. The combination of chip-scale Kerr frequency comb sources with large-scale silicon photonic integration could hence become a key concept for power-efficient optical interconnects providing transmission rates that have hitherto been considered impossible.

The on-chip pump power needed for Kerr comb generation discussed in Section 5.2.3, still amounts to 29 dBm. However, there is considerable room for future improvements – Kerr combs have been demonstrated with a threshold power as low as 50 µW in silica toroids [66]. In general, the threshold for Kerr comb generation scales inversely quadratic with the Q-factor of the resonator [72]. A 10-fold Q-factor improvement would lead to a reduction of the threshold pump power by a factor of 100, and the 3 W pump laser could be replaced by a 30 mW pump laser diode [66].

Although solving the problem of spectral flatness, soliton comb spectra are typically characterized by a bad conversion efficiency [180], and the broad spectrum exceeds the telecommunication bands such that many comb lines are not useful for data transmission with current technology for transmitters, amplifiers and receivers. Further studies towards spectral shaping of Kerr frequency combs are required to optimize the spectral envelope, coherence properties and conversion efficiency to fully leverage the high potential of Kerr combs for coherent data transmission. A new comb state that has been named 'Platicon' has been predicted in [181]. Platicons show two advantages as compared to solitons, i.e., higher conversion efficiency and a spectral envelope that exhibits more lines with similar power in the center and a steeper roll-off to the edges of the spectrum.

6 Summary and future work

6.1 Summary

This chapter provides a summary of the comb generation concepts that have been investigated in this thesis and compares them with other state-of-the-art comb-based data transmission experiments. In addition, potential improvements of each comb generation technology are discussed. This section is an extension to the comparison that was done in Section 5 of [J2].

Apart from the comb sources investigated in this thesis, Tbit/s super-channel transmission has been demonstrated using different approaches for realizing the optical source. These approaches include ensembles of independent lasers [127], [128], single-laser concepts with sidebands generated by external modulation [51], [129], and fiber-based spectral broadening [50], [52], [53], [113], [130].

Arrays of independent lasers offer flexibility and do not require filters for spectrally separating the comb lines. However, the achievable spectral efficiency (SE) is limited by the drift of the individual emission wavelengths and the necessary spectral guard bands.

For modulator-based comb sources, the number of comb lines is generally limited by the achievable modulation depth [J5] unless complex arrangements of cascaded modulators with synchronized driving signals are used [101], [129], [131]. Moreover, if a large number of lines is to be generated, the power of the comb lines is limited by the power handling capacity and by the insertion loss of the modulator [102].

Broadband frequency combs for Tbit/s super-channels can also be generated by solid-state mode-locked lasers in combination with highly nonlinear fibers [50], [52], [112], or multi-stage parametric mixers [53], [113], [130]. However, both approaches require strong optical pumps and large interaction lengths in delicately arranged sequences of specialized optical fibers. Integration density is hence limited.

Table 6-1 summarizes key parameters of the experiments presented in this thesis, and compares them with other state-of-the-art experiments. In lines 1 – 10 of the table, the results presented in Chapters 3 to 5 are listed. Lines 11 and 12 represent other experiments performed in the same lab, and lines 13 to 15 show data from literature. The specified values relate only to those comb lines that were actually used for the respective data transmission experiment. The ranges refer to the minimum and maximum quantities, respectively.

Table 6-1: Comparison of frequency comb generators for Tbit/s data transmission

	Comb generator type	FSR [GHz]	# of lines used	Power per line [dBm]	OCN_0R [dB Hz]	Aggr. net data rate [Tbit/s]	Ref.
1	GSCS	20	13	–28.2 … –7.1	117 … 129	0.9	[J2]
2	GSCS	20	8	–10.7 … –7.1	126 … 128	1.0	[J2]
3	GSCS	18.5	15	–25.7 … –8.0	119 … 132	1.0	[J2]
4	GSCS	18.5	9	–11.4 … –8.0	122 … 132	1.1	[J2]
5	GSCS	12.5	24	–32.0 … –8.9	114 … 129	1.1	[J2]
6	GSCS	12.5	24	–32.0 … –8.9	114 … 129	1.9	[J2]
7	QD-MLLD	42	29	–13.9 … –7.2	143 … 146	1.0	[C6]
8	QD-MLLD	42	25	–13.2 … –7.4	> 136	1.6	[C7]
9	Kerr comb	25	20	–32.3 … –5.2	112 … 139	1.3	[J4]
10	Kerr soliton	95.8	94	–24.1 … –13.3	> 158	19.7	[C5]
11	SOH modulator	25	9	–34.7 …–17.1	117 … 135	1.1	[J5]
12	MLL+HNLF	12.5	325	N.A.	119 … 141	26.0	[52]
13	Casc. $LiNbO_3$ Mod	6.48	20	N.A.	N.A.	1.1	[129]
14	2 Mod in RFS	25	112	~ –15…0	> 120	10.0	[51]
15	Parametric comb	6.25	1520	N.A.	~120	29.7*	[53]

* Estimated value, only few channels measured
N.A.: Not available

The QD-MLLD as well as the central lines of the GSCS show the best comb uniformity at an acceptable average power level, which is only surpassed by a few lines of the Kerr comb and the recirculating frequency shifter (RFS). With respect to the OCN_0R, the best values are found for the soliton Kerr comb. It has to be noted that all comb sources listed in Table 6-1, lines 1 to 10, are not yet fully optimized. Further steps towards a hybrid or even a monolithic integration are currently being investigated, and performance improvements are to be expected for all comb generation concepts.

For the case of the GSCS, lines 1 to 6, it is possible to monolithically integrate two lasers, where the master laser synchronizes the slave laser by injecting light at the back facet of the slave, while the comb is collected from its front facet [135]. This removes the need for a circulator and a polarization controller as in Fig. 3.1.

QD-MLLD, lines 7 and 8, are the most compact comb sources, however, are very sensitive towards external perturbations, noise of the current source and the TEC as well as optical feedback. For coherent data transmission the limiting factor is the phase noise of the comb lines. Line 7 summarizes the results achieved with the feed-forward phase noise compensation. This, however, comes at the price of reduced power and spurious side-lines. Subsequent filtering and amplification is required, which reduces the OSNR of the data channels. Line 8 shows the values for the experiment using self-homodyne detection, which is tolerant towards phase noise, however, reduces spectral efficiency as only a single polarization is used for the data.

The Kerr frequency comb listed in line 9 of Table 6-1 suffers from large line-to-line power variations, which limit the number of available channels and hence the achievable data rate. This limitation can be overcome by generating so-called soliton combs as described in Section 5.2.4. Soliton combs were demonstrated with crystalline resonators [71] and with silicon-nitride resonators [182]. Using such a soliton Kerr comb allows to increase the data rate to 19.7 Tbit/s, see line 10 of Table 6-1. Even higher data rates are possible as the limitations were set by the available equipment in the laboratory. Further advances in the fabrication processes will also help reducing the pump power.

The SOH comb generator, line 11, is a promising approach, since it also allows for easy adjustment of center frequency and line spacing of the comb. Moreover, the SOH modulators which generate the comb feature very low U_π and can thus produce a reasonably large number of comb lines with moderate RF driving powers. It has been shown in [J5] that a dual-drive SOH modulator can be used to generate a flat comb with seven lines within 2 dB, whereas the data transmission results listed here still stem from a single-drive modulator. As with any modulator-based approach, it is possible to achieve a higher number of lines either by cascading multiple modulators [129] or by placing the modulator(s) into a loop [51].

Lines 13 and 14 of Table 6-1 show the results obtained for frequency comb generation based on cascaded modulators. The former experiment relies on a multi-core transmission fiber and demonstrates net data rates of 1.12 Tbit/s per core by using orthogonal frequency division multiplexing. The latter experiment used two cascaded phase modulators in a RFS and shows transmission of an OFDM data stream over 640 km SSMF with a net capacity of 10 Tbit/s.

For the cases of lines 12 and 15, the experiments were conducted using parametric comb generators that exploit Kerr nonlinearities in optical fibers, cf. [53] and [52], respectively. Using two interleaved 12.5 GHz combs generated by multi-stage parametric mixing, data transmission on 77 carriers was demonstrated. Extrapolating thc performance of these carriers to the ensemble of 1520 carriers, a total data rate of 31.8 Tbit/s was estimated. Previously, transmission of 32.5 Tbit/s (net data rate 26 Tbit/s, see line 12 of Table 6-1) was demonstrated using 325 carriers generated by a solid state mode-locked laser along with spectral broadening in a highly nonlinear fiber. These experiments demonstrate the potential of parametric frequency comb generation in conventional systems consisting of discrete elements. This performance can be also achieved with chip-scale systems, e.g., with soliton Kerr frequency combs generated in a microresonator [C5].

6.2 Outlook

With the exception of the on-off-keying experiments using a Kerr comb source and the self-homodyne experiments with a QD-MLLD all data transmission experiments used a coherent receiver with an external cavity laser (ECL) as a local oscillator. To fully leverage the potential of frequency combs, a second comb should be used as a multi-wavelength local oscillator (LO). In this case, only two parameters need to be controlled to fully synchronize the local oscillator comb to the transmitter comb rather than having to control each LO individually. A first proof-of-principle experiment has been performed using two GSCS and a two-channel digital control loop for synchronization [C1], demonstrating a Tbit/s data transmission over 10 km. Further investigations are required to understand the limitations when using a frequency comb as a local oscillator in order to design an optimized frequency comb based transceiver.

Future investigations towards comb-based Tbit/s transceivers could focus on specific application areas and optimize the comb sources and system implementations according to the respective requirements in terms of size, cost, power consumption, reach, spectral efficiency and data rate.

For the first generation of Tbit/s transceivers the GSCS seem best suited as they provide a sufficient number of lines and are currently the most mature technique, allowing for the shortest time-to-market. A second generation of Tbit/s transceivers might be using QD-MLLD comb sources, which could reduce power consumption compared to the GSCS as there are no high-frequency electronics involved. Apart from the two demonstrated options for mitigating the influence of phase noise, an increase of the symbol rate is an option. As discussed in Section 2.2.2, the tolerance towards phase noise increases with the symbol rate. For devices with an FSR of 42 GHz, a symbol rate of 40 GBd can be employed. This should allow coherent data transmission without the feed-forward phase noise compensation and provides a path towards a multi-Tbit/s transceiver. If 30 lines could be used with PDM-QPSK at 40 GBd, the resulting line rate would be 4.8 Tbit/s. Even higher data rates can be achieved with Kerr comb sources. However, in order to make these devices suitable for communication systems there are several improvements necessary, see discussion in Section 5.3.

Appendix A Nonlinear microresonators

As a simple example we discuss a ring resonator. The resonance condition in the linear case will be established, and an expression for the frequency-dependent free spectral range (FSR) will be derived. The non-equidistant spacing of the resonance frequencies has important consequences for the Kerr frequency comb generation in microresonators. Non-linear effects leading to Kerr comb formation inside the microresonator will be introduced. Finally we discuss how solitons in microresonators can be generated. These so-called soliton combs are preferred for optical data transmission.

A.1 Resonance condition and dispersion

In this section we consider a resonator that is formed using a dielectric waveguide. The waveguide is assumed to be transversely single-moded. We introduce the resonance condition and derive the longitudinal mode spacing.

We consider a simple ring resonator as depicted in Fig. A.1. The same equations that will be derived throughout this chapter also apply to any arbitrary resonator geometry with roundtrip length L. For the case of a Fabry-Perot resonator that means a mirror separation of $L/2$.

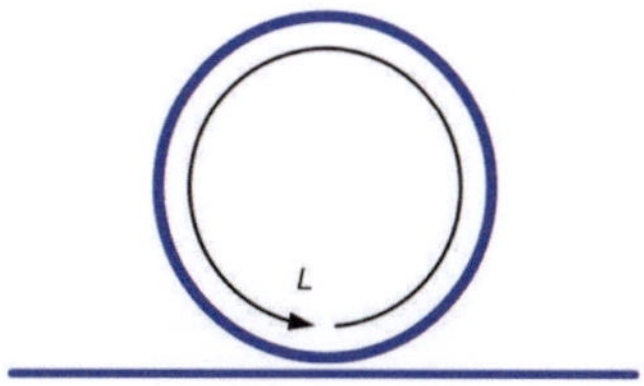

Fig. A.1: Ring resonator with circumference L and an evanescently coupled waveguide.

A resonant field inside the resonator can only exist if its amplitude and phase reproduce after one round-trip. With the propagation constant $\beta(\omega)$ at angular frequency ω and the circumference L of the ring, the resonance condition can be formulated as

$$\beta(\omega)L = \mu \cdot 2\pi \, . \tag{A.6}$$

The quantity μ is an integer. The propagation constant $\beta(\omega)$ is given by

$$\beta(\omega) = n_e(\omega)\frac{\omega}{c_0}, \tag{A.7}$$

where $n_e(\omega)$ is the effective index of the transverse mode guided by the resonator waveguide, and c_0 is the vacuum speed of light.

The resonance angular frequencies $\omega_\mu = \omega(\mu)$ around a central resonance angular frequency $\omega_c = \omega(0)$ can be approximated by a Taylor series[15] in the integer mode index μ, which is regarded as a continuous variable for the moment,

$$\begin{aligned} \omega(\mu) &= \omega_c + D_c^{(1)}\mu + \frac{1}{2}D_c^{(2)}\mu^2 + \ldots \\ &= \omega_c + \sum_{k=1}^{\infty}\frac{1}{k!}D_c^{(k)}\mu^k, \qquad D_c^{(k)} = \left.\frac{\partial^k \omega}{\partial \mu^k}\right|_{\mu=0} . \end{aligned} \tag{A.8}$$

To evaluate the coefficients $D_c^{(k)}$ the following auxiliary calculation is helpful, which uses Eqs. (A.6), (A.7), and the definition of the effective group refractive index $n_{eg}(\omega) = n_e(\omega) + \omega \partial n_e(\omega)/\partial\omega$,

$$\frac{\partial \omega n_e(\omega)}{\partial \mu} = n_e(\omega)\frac{\partial \omega}{\partial \mu} + \omega\frac{\partial n_e(\omega)}{\partial \omega}\frac{\partial \omega}{\partial \mu} = n_{eg}(\omega)\frac{\partial \omega}{\partial \mu} = \frac{2\pi c_0}{L}. \tag{A.9}$$

Recalling the discrete nature of the mode index μ yields $\partial\mu \to \Delta\mu = 1$, so that $\partial\omega \to \Delta\omega = 2\pi c_0 / n_{eg}(\omega)L$ results as the FSR. The first coefficient $D_c^{(1)}$ corresponds to the FSR at the central resonance and describes an equidistant grid $\omega_\mu = \omega_c + D_c^{(1)}\mu$. Higher-order coefficients $D_c^{(k>1)}$ are a measure for the deviation of the resonance frequencies from the equidistant grid.

The coefficients $D_c^{(k)}$ are related to the coefficients $\beta_c^{(i)}$, which are used in Section B.3.2 to describe dispersion in waveguides,

[15] It is assumed that $\omega(\mu)$ is an analytic function.

$$D_c^{(1)} = \left.\frac{\partial \omega}{\partial \mu}\right|_{\mu=0} = \frac{2\pi c_0}{L n_{eg}(\omega_c)} = \frac{2\pi}{L}\frac{1}{\beta_c^{(1)}},$$
$$D_c^{(2)} = \left.\frac{\partial^2 \omega}{\partial \mu^2}\right|_{\mu=0} = -\left(\frac{2\pi c_0}{L}\right)^2 \left.\frac{\partial n_{eg}(\omega)/\partial \omega}{n_{eg}^3(\omega)}\right|_{\mu=0} = -\left(\frac{2\pi}{L}\right)^2 \frac{\beta_c^{(2)}}{\left(\beta_c^{(1)}\right)^3}. \quad \text{(A.10)}$$

For normal dispersion, $D_c^{(2)} < 0$ and $\beta_c^{(2)} > 0$ holds, while for anomalous dispersion, $D_c^{(2)} > 0$ and $\beta_c^{(2)} < 0$ is true. In the next section, we show that dispersion plays a critical role for comb generation in microresonators.

A.2 Nonlinear processes leading to Kerr comb generation

In nonlinear optics the term "nonlinear" refers to the response of the polarization $\mathbf{P}$ of a dielectric material to an applied electric field $\mathbf{E}$, which becomes nonlinear if the field is particularly high. For simplicity, scalar quantities for the polarization $P(t)$ and the electric field $E(t)$ are assumed. The polarization can then be written as a power series [107]

$$P(t) = \varepsilon_0 \chi^{(1)} E(t) + \varepsilon_0 \chi^{(2)} E^2(t) + \varepsilon_0 \chi^{(3)} E^3(t) + \ldots, \quad \text{(A.11)}$$

where ε_0 is the vacuum permittivity and $\chi^{(n)}$ are the n-th order susceptibilities. Equation (A.11) does not only neglect the vectorial nature of electrical field and polarization, it also assumes that the nonlinear medium is time invariant, local in time and space, isotropic and homogeneous. Time locality means that the medium needs to be lossless and dispersionless [107]. Still, Eq. (A.11) is helpful to obtain a qualitative understanding of nonlinear optics.

The following discussion is limited to third-order optical nonlinear effects, which are relevant for Kerr comb generation, i.e., to self-phase modulation (SPM), cross-phase modulation (XPM), and four-wave mixing (FWM). In order to understand these effects, it is instructive to calculate the third-order polarization $P^{(3)}(z,t) = \varepsilon_0 \chi^{(3)} E^3(z,t)$ for the case of the superposition of three plane waves with three different frequencies $f_l = \omega_l / 2\pi$, $l = 1 \ldots 3$ that co-propagate in z-direction. The electric field $E(z,t)$ is described using analytic complex time-domain signals

$$E(z,t)=\sum_{l=1}^{3}\operatorname{Re}\left[\underline{E}(z,\omega_l)\mathrm{e}^{\mathrm{j}\omega_l t}\right]=\sum_{l=1}^{3}\frac{1}{2}\left(\underline{E}_l\mathrm{e}^{\mathrm{j}(\omega_l t-k_l z)}+\underline{E}_l^{*}\mathrm{e}^{-\mathrm{j}(\omega_l t-k_l z)}\right). \quad (A.12)$$

The nonlinear polarization for the specific effects of interest can now be calculated by taking the third power of Eq. (A.12) and by sorting the terms.

Self-phase modulation (SPM):

$$P_{\mathrm{NL}}(z,t)=\frac{1}{2}\epsilon_0\chi^{(3)}\left(\frac{3}{4}\left|\underline{E}_1\right|^2\underline{E}_1\mathrm{e}^{\mathrm{j}(\omega_1 t-k_1 z)}+\ldots+\mathrm{c.c.}\right) \quad (A.13)$$

Cross-phase modulation (XPM):

$$P_{\mathrm{NL}}(z,t)=\frac{1}{2}\epsilon_0\chi^{(3)}\left(\frac{6}{4}\left|\underline{E}_2\right|^2\underline{E}_1\mathrm{e}^{\mathrm{j}(\omega_1 t-k_1 z)}+\ldots+\mathrm{c.c.}\right) \quad (A.14)$$

Four-wave mixing (FWM):

$$\begin{aligned}P_{\mathrm{NL}}(z,t)=\frac{1}{2}\epsilon_0\chi^{(3)}\Bigg(&\frac{3}{4}\left|\underline{E}_1\right|^2\underline{E}_2^{*}\mathrm{e}^{\mathrm{j}((2\omega_1-\omega_2)t-(2k_1-k_2)z)}+\ldots+\mathrm{c.c.}\\&+\frac{6}{4}\underline{E}_1\underline{E}_2\underline{E}_3^{*}\mathrm{e}^{\mathrm{j}((\omega_1+\omega_2-\omega_3)t-(k_1+k_2-k_3)z)}+\ldots+\mathrm{c.c.}\Bigg)\end{aligned} \quad (A.15)$$

In these equations, the "+…" comprises all terms with all permutations of the subscripts, and c.c. stands for the complex conjugate of the previous terms. From Eqs. (A.13) and (A.14), it can be seen that the polarization oscillates at the same frequency as the input field and that it is proportional to the magnitude square of either the input field at the same frequency (SPM) or another frequency (XPM). Contrary, the case of FWM, Eq. (A.15), shows that the polarization actually oscillates at a frequency different from the frequencies of the input fields, i.e., new frequencies are generated. While SPM and XPM are inherently phase matched, FWM requires additional measures for phase matching.

As introduced in Section 2.1.2, the Kerr comb is generated by pumping a single resonance with a continuous wave (CW) laser. The underlying process is degenerate FWM, i.e., all terms in Eq. (A.15) that contain two identical frequencies (e.g. $2\omega_1-\omega_2$). These processes require both energy conservation and phase matching. As neighboring resonances are required to have a fixed phase difference of 2π, see Eq. (A.6), phase matching is inherently fulfilled for any combination of resonances symmetrically located to the right and to the left of a central resonance. We denote these three resonances with the indices c and $\pm k$. Fig. A.2(a) shows the dispersion diagram for the cold cavity case, i.e., in the absence of any pump power. The solid blue line marks the case for anomalous dispersion. It can be seen that the equidistant β lead to non-equidistant resonance frequencies. Thus the condition $2\omega_c = \omega_k + \omega_{-k}$ cannot be fulfilled. When pumping the resonance, however, the additional SPM and XPM lead to shifts of the resonance frequencies as shown in Fig. A.2(b). As XPM is twice as strong as SPM, which can be seen from the degeneracy factors in Eqs. (A.13) and (A.14), it is possible to find a situation where the three resonances are equidistant.

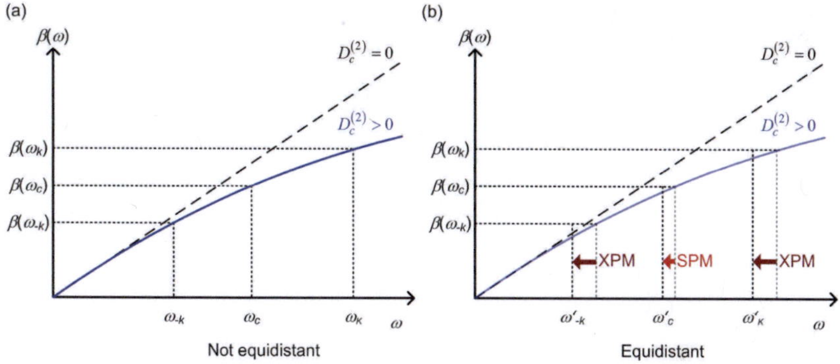

Fig. A.2: Dispersion diagram for a resonator with and without nonlinear shifts of the resonance frequencies. (a) Dispersion diagram for the cold cavity. The dashed line shows the case without group velocity dispersion, while the solid blue line shows the case for anomalous dispersion. The resonance condition requires equidistant $\beta(\omega_\mu)$, hence the corresponding resonance frequencies ω_μ differ from equidistance, which is depicted here for two resonances symmetrically located around a central resonance, i.e., $\mu = c$ and $\mu = \pm k$. (b) Shift of the resonance frequencies when the resonance at ω_c is pumped. This resonance experiences a phase shift due to SPM, which is only half of the phase shift experienced by all other resonances due to XPM. It is thus possible to find three equidistant resonances, to initiate comb generation by FWM.

This situation is sometimes compared with modulation instability in an optical fiber [183], [184]. Modulation instability is observed when a strong CW laser is transmitted via a fiber with anomalous dispersion. A modulated signal is observed after some propagation distance. This is due to sidebands generated by FWM.

A.3 Soliton generation inside a microresonator

If a waveguide exhibits anomalous group velocity dispersion then the effects of dispersion and self-phase modulation can cancel. Both, the dispersion and the self-phase modulation cause a chirp in a propagating pulse. For the case of anomalous dispersion the induced chirp is of opposite sign than for the self-phase modulation. Therefore, it can be shown that certain pulse shapes exist which propagate along this waveguide unchanged. By choosing the correct peak power along with the right duration of the pulse, the dispersion-induced chirp can be completely canceled [106]. This waveform is then called a soliton. For a mathematical treatment of soliton propagation in waveguides, refer to [106], [107]. For soliton propagation in microresonators, see the Supplementary Information of [71].

A practical challenge for generating and maintaining a soliton in a microresonator stems from the thermal dynamics of the resonances when being pumped. Due to the absorption of pump light, the resonator heats up, which affects the resonance frequencies in two ways [185]. First, the physical length of the resonator changes due to thermal expansion of the materials, and second, the propagation constant changes due to the temperature dependence of the refractive indices of the materials.

In the following, we denote with "ring resonance" the "cold" ring resonance closest to the pump frequency. If the CW pump laser frequency is tuned into a ring resonance from the higher frequency side, then increasingly more pump power is absorbed. As a consequence, the ring heats up so that the refractive index increases, and the resonance frequency decreases, i.e., it "runs away" from the pump. We consider a ring resonator with an evanescently coupled bus waveguide as depicted in Fig. A.1, and measure the transmitted power at the bus output. With decreasing pump frequency and increased resonant absorption, the bus power decreases. If for a given pump power the pump frequency is down-tuned beyond the "hot" resonator center frequency, the heating cannot be increased any more, and the resonance frequency cannot become larger. As soon as the pump laser leaves the maximum of the microring resonance, the resonator cools down and the resonance shifts back to its original frequency so that the bus power jumps back to its original high level. Within this range of decreasing bus power, the resonance is thermally locked [185] to the pump laser. Thermal locking by a pump detuning towards higher frequencies is not possible, because the resonance “runs towards” the pump laser frequency, which prevents an equilibrium state.

The formation of a Kerr comb from a single pump line is described in Fig. 5.1. Thermal locking fixes the ring’s resonance frequency to the low-frequency side of the pump frequency. In the time domain, a frequency comb with a regular line spacing f_r corresponds to an impulse which travels in the microresonator with a round-trip time (pulse repetition rate) of $\tau_r = 1/f_r$. If the pump power inside the ring is further increased by down-tuning the pump frequency, the pulse power increases and reaches eventually the soliton threshold, where the initital pulse travelling round in the microresonator takes the shape of a soliton through the interplay between a Kerr-induced increase of the refractive index with intensity and anomalous dispersion. In the soliton state, the ring resonance is slightly shifted to the high frequency side of the pump line [71], and thermal locking as described above does not occur. Therefore, the soliton state is reached stably only if the tuning speed of the pump frequency and the temperature response of the resonance are appropriately chosen: If the tuning speed is much faster than the resonator’s thermal response, sufficient pump power does not build up in the ring. If the tuning speed is much smaller than the ring’s thermal response, no stable state is reached for pump frequencies smaller than the actual resonance.

If a soliton comb has been established, the pump power as coupled to the ring must supply only the soliton power which is coupled out to the bus. Therefore, the ring temperature decreases and the resonance frequencies approach their “cold” values.

Appendix B Optical data transmission

In this section, we provide the necessary background for data modulation, fiber transmission and data reception as required for this thesis. The discussion is limited to direct-detection and coherent data transmission over standard-single mode fiber (SSMF) using polarization and wavelength-division multiplexing.

B.1 Transmitter

The following paragraphs describe the basic concepts of digital modulation, and the physical means to encode digital data onto an optical carrier.

B.1.1 Modulating digital data onto an optical carrier

A binary data stream consists of logical '0's and logical '1's. In order to transport these bits from one location to another, it is necessary to convert the bits into analog waveforms. Easiest to understand is to use the presence of a light pulse as a '1' and the absence of a pulse as a '0', which is commonly referred to as on-off keying (OOK).

Data can be transmitted more efficiently if it is possible to distinguish between several different waveforms. In case M distinct waveforms are used, then $\log_2 M$ bits can be encoded onto each of them. One symbol encodes $\log_2 M$ bits, and the complete set of M symbols is called the symbol alphabet. Modulation formats are distinguished by the arrangement of the symbol alphabet in an n-dimensional space, where n is the number of independent degrees of freedom to define the physical waveform.

An optical waveform is fully described by its amplitude, phase, frequency and polarization. In most practical cases, and throughout this thesis, frequency and polarization are used for multiplexing, see Section B.2. This leaves two degrees of freedom to define the analog waveforms for transmission, the amplitude $A(t)$ and the phase $\varphi(t)$ of the optical carrier with central frequency $\omega_c = 2\pi f_c$. The associated optical electric field $\underline{E}(t)$ can be expressed as an analytical signal

$$\underline{E}(t) = A(t)\mathrm{e}^{\mathrm{j}\varphi(t)}\mathrm{e}^{\mathrm{j}\omega_c t}. \tag{B.1}$$

B.1.2 Modulation formats

As we use two degrees of freedom, it is convenient to represent and compare modulation formats in two-dimensional space using the phasor $\underline{S}(t) = A(t)\exp(\mathrm{j}\varphi(t))$. This phasor can be visualized in the complex plane. A display of symbols in this plane is named a constellation diagram or an IQ diagram, see Fig. B.1. The 'I' stands for the inphase (real part), and the 'Q' for the quadrature component (imaginary part of the electric field).

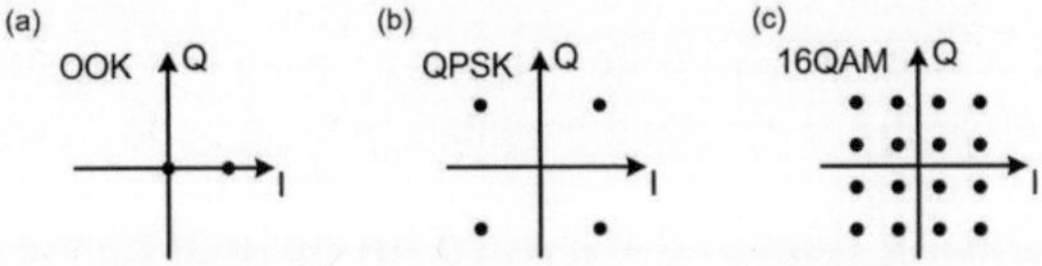

Fig. B.1: Constellation diagrams of popular modulation formats. (a) On-off keying (OOK). (b) Quadrature phase shift keying (QPSK). (c) 16-state quadrature amplitude modulation (16QAM).

Fig. B.1(a) shows the constellation diagram for on-off keying (OOK), which uses two symbols and encodes one bit per symbol. Four symbols, each representing two bits, are used in the case of quadrature phase shift keying (QPSK), Fig. B.1(b). A number of four bits per symbol is used for 16-state quadrature amplitude modulation (16QAM), Fig. B.1(c).

Physically it is of course neither possible to transmit complex signals nor to measure them at the receiver. However, it is possible to transmit two independent data streams and interpret these as the real and imaginary part of a complex number. Taking the real part $E_r(t) = \mathrm{Re}\left[\underline{E}(t)\right]$ leads to the definition of two independent data signals, the in-phase signal $I(t)$ and the quadrature signal $Q(t)$

$$\begin{aligned} E_r(t) &= \mathrm{Re}\left[\underline{E}(t)\right] = A(t)\cos(\omega_c t + \varphi(t)) \\ &= A(t)\cos(\varphi(t))\cos(\omega_c t) - A(t)\sin(\varphi(t))\sin(\omega_c t) \\ &= I(t)\cos(\omega_c t) - Q(t)\sin(\omega_c t). \end{aligned} \tag{B.2}$$

It can be seen from Eq. (B.2) that the physical way to transmit complex data signals is to mix the in-phase component $I(t)$ with a carrier $\cos(\omega_c t)$ and to mix the quadrature component $Q(t)$ with an orthogonal carrier $\sin(\omega_c t)$.

Comparing different constellation diagrams, one important practical aspect is that the larger the number of constellation points is, the smaller their

Euclidian distance in constellation space becomes[16], see Fig. B.1. This results in a lower noise tolerance, see Fig. 2.4 in Section 2.2.1, where it is shown that 16QAM needs a higher optical signal-to-noise power ratio (OSNR) than QPSK to achieve the same bit-error-ratio (BER).

B.1.3 Electro-optical modulator

There are several ways to control the optical waveform with an electrical signal. The simplest one is to directly modulate the current of a semiconductor laser, thereby changing its output power. Alternatively, the optical power can be modulated using an electro-absorption modulator. This device is based on the Franz-Keldysh effect, where the absorption of a semiconductor is changed by applying an electric modulation field. The following discussion concentrates on the Mach-Zehnder modulator (MZM), which can be used as an amplitude modulator or as a phase modulator. This allows generating the constellation diagrams as depicted in Fig. B.1.

A MZM is a Mach-Zehnder interferometer with phase shifters in each arm as depicted in Fig. B.2(a). The electric field at the input is split in two paths. Depending on the phase relation of the light in the two arms of the interferometer, there can be constructive or destructive interference at the output. This phase relation is set by two phase shifters, which can be externally controlled. The phase shifters typically exploit the electro-optic Pockels effect, where the refractive index of a material changes in proportion to the applied electric field strength. Fig. B.2 depicts an MZM with three phase shifters. Two of them are used for encoding the time-dependent data voltages, $u_1(t)$ and $u_2(t)$. The third one is used to apply a constant bias voltage u_{DC}, which sets the desired operating point. The induced phase shifts are denoted by $\Delta\phi_1(t)$, $\Delta\phi_2(t)$, and $\Delta\phi_{DC}$, respectively.

[16] If the average power remains constant.

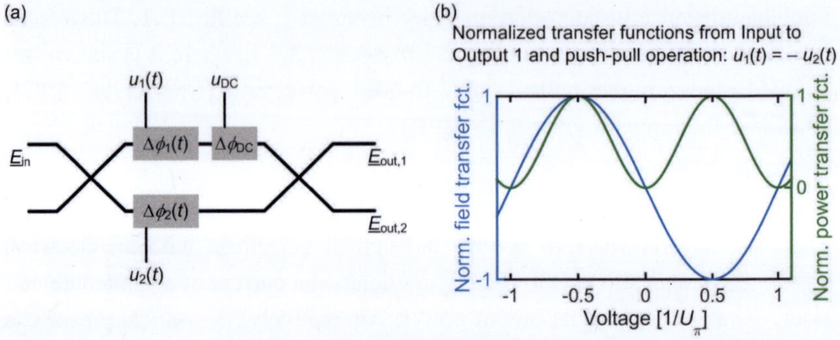

Fig. B.2: Mach-Zehnder modulator (MZM). (a) Schematic of a MZM. It consists of a Mach-Zehnder interferometer with phase shifters in each arm. For modulating the optical field, the phase shifters are controlled by three voltages, two time dependent signals $u_1(t)$ and $u_2(t)$ and a DC bias u_{DC}. The induced phase shifts are given by $\Delta\phi_1(t)$, $\Delta\phi_2(t)$ and $\Delta\phi_{\mathrm{DC}}$, respectively. (b) Normalized optical electric field amplitude (left, blue axis) and optical power (green, right axis) transfer function from the input to output 1 for the case of push-pull operation, i.e., for $u_1(t) = -u_2(t)$.

The transfer functions for the two outputs fields $\underline{E}_{\mathrm{out},1}$ and $\underline{E}_{\mathrm{out},2}$ relative to the input field $\underline{E}_{\mathrm{in}}$, the so-called optical amplitude transfer functions, can be written as

$$\frac{\underline{E}_{\mathrm{out},1}}{\underline{E}_{\mathrm{in}}} = \mathrm{j}\mathrm{e}^{\mathrm{j}\frac{\Delta\phi_1(t)+\Delta\phi_2(t)}{2}+\mathrm{j}\frac{\Delta\phi_{\mathrm{DC}}}{2}} \sin\left(\frac{\Delta\phi_1(t)-\Delta\phi_2(t)}{2}+\frac{\Delta\phi_{\mathrm{DC}}}{2}\right) \tag{B.3}$$

and

$$\frac{\underline{E}_{\mathrm{out},2}}{\underline{E}_{\mathrm{in}}} = \mathrm{j}\mathrm{e}^{\mathrm{j}\frac{\Delta\phi_1(t)+\Delta\phi_2(t)}{2}+\mathrm{j}\frac{\Delta\phi_{\mathrm{DC}}}{2}} \cos\left(\frac{\Delta\phi_1(t)-\Delta\phi_2(t)}{2}+\frac{\Delta\phi_{\mathrm{DC}}}{2}\right). \tag{B.4}$$

For the special case that the upper arm is driven with the negative signal of the lower arm, i.e., $\Delta\phi_2(t) = -\Delta\phi_1(t)$, the transfer functions simplify to

$$\frac{\underline{E}_{\mathrm{out},1}}{\underline{E}_{\mathrm{in}}} = \mathrm{j}\mathrm{e}^{\mathrm{j}\frac{\Delta\phi_{\mathrm{DC}}}{2}} \sin\left(\frac{\Delta\phi(t)}{2}+\frac{\Delta\phi_{\mathrm{DC}}}{2}\right) \tag{B.5}$$

and

$$\frac{\underline{E}_{\mathrm{out},2}}{\underline{E}_{\mathrm{in}}} = \mathrm{j}\mathrm{e}^{\mathrm{j}\frac{\Delta\phi_{\mathrm{DC}}}{2}} \cos\left(\frac{\Delta\phi(t)}{2}+\frac{\Delta\phi_{\mathrm{DC}}}{2}\right), \tag{B.6}$$

respectively, where $\Delta\phi(t)=\Delta\phi_1(t)-\Delta\phi_2(t)=2\Delta\phi_1(t)=-2\Delta\phi_2(t)$. The power transfer function can be obtained by taking the magnitude squared of Eq. (B.3)-(B.6), respectively. The amplitude and power transfer functions for this so-called push-pull operation are depicted in Fig. B.2(b).

For using such a MZM to generate OOK signals, the bias voltage is typically adjusted in the center between the maximum and the minimum power transmission where the power slope is maximal. This is called the quadrature point. For modulating complex data signals, the MZM is biased at the zero of the amplitude transfer function, where the amplitude slope is maximal. As this corresponds to zero power transmission, it is also called the null-point of the modulator.

In order to generate IQ signals, two MZM are nested into a Mach-Zehnder structure with an additional 90° phase shift for one of the arms, see Fig. B.3. One arm of the outer Mach-Zehnder interferometer is used to modulate along the I-axis, while the other one modulates along the Q-axis. This is exemplarily indicated in Fig. B.3 for the case of 16QAM.

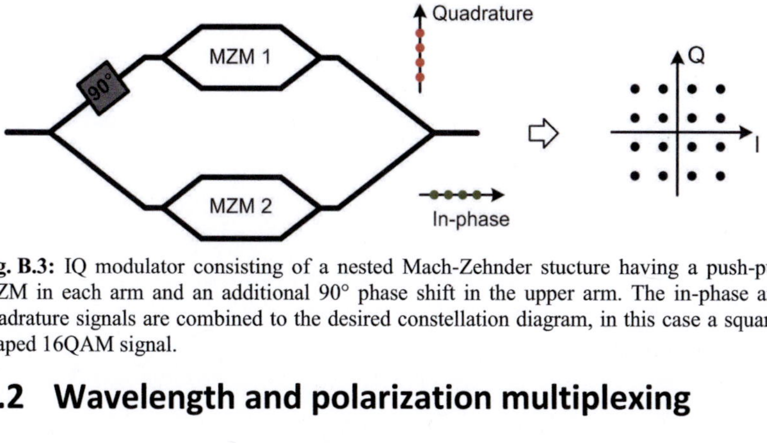

Fig. B.3: IQ modulator consisting of a nested Mach-Zehnder stucture having a push-pull MZM in each arm and an additional 90° phase shift in the upper arm. The in-phase and quadrature signals are combined to the desired constellation diagram, in this case a square-shaped 16QAM signal.

B.2 Wavelength and polarization multiplexing

The optical transmission fiber is generally operated in its linear regime by keeping the optical input power low compared to the threshold for nonlinear effects. In this case, optical waveforms superimpose without affecting each other. Therefore waveforms can be sent in parallel if it is possible to distinguish between them at the receiver. The most commonly used technique is the so-called wavelength-division multiplexing (WDM), where different laser wavelengths, i.e., different optical carrier frequencies are used, which can be separated at the receiver by optical, electrical or digital band-pass filters.

A crucial metric for WDM systems is the spectral efficiency (SE), which is the capacity C in bit/s devided by the bandwidth B of the WDM channel [7], [186], [187]. To investigate an upper limit for the SE, we assume that the signal is using sinc-pulses in time domain with a symbol period T. The associated bandwidth is given by $1/T$, which is equal to the symbol rate R_s and is less than B. Further we assume that a unit-energy matched-filter is used at the receiver and that a white Gaussian noise distribution [7] is true. The spectral efficiency can then be expressed as a function of the signal-to-noise power ratio (SNR)

$$\mathrm{SE} = \frac{C}{B} = \frac{R_s}{B} \log_2\left(1 + \mathrm{SNR}\right) \tag{B.7}$$

Note that the SNR also depends on the symbol rate. The noise power, which is in the denominator of the SNR is integrated over the receiver bandwidth that scales with the symbol rate. Therefore, the SNR will decrease for increasing symbol rate assuming that the signal power in the numerator of the SNR remains constant.

The advantage of using sinc-pulses rather than traditional non-return-to-zero (NRZ) signaling is illustrated in Fig. B.4. In panels (a) to (c), the upper row shows the NRZ-case, while the lower row depicts the same physical quantity for sinc pulse shaping. Panel (a) of Fig. B.4 shows the time-domain signal for a sequence of five bits. For the case of NRZ this means a high (low) signal level during the entire respective bit slot. In contrast, the sinc pulses are not confined to a single bit slot. The width of the sinc pulses is chosen such that the maximum of one impulse coincides with the zeros of all other impulses. At the center of each bit slot there is hence no intersymbol interference.

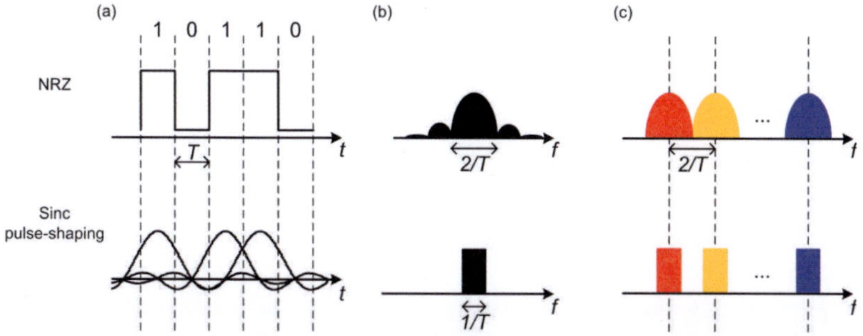

Fig. B.4: Wavelength-division multiplexing (WDM) without pulse shaping (upper row) and with pulse shaping (lower row). (a) Time signals for a bit sequence of '10110' for the case of non-return-to-zero (NRZ) and with sinc-shaping (b) Corresponding single channel power spectra. The power spectrum of the NRZ signal has the shape of a sinc squared. For the sinc-pulse shaping all spectral components reside within the Nyquist bandwidth. (c) Packing of WDM channels on a regular grid given by $2/T$. Side lobes of the sinc can be suppressed by optical filters prior to multiplexing the WDM channels. It can be seen that for the case of sinc pulse-shaping the channels could either be packed much denser or the symbol rate could be increased.

In the frequency domain, Fig. B.4(b), the NRZ signal has the shape of a sinc-function, here represented by its power spectral density. In practice, the side lobes of the spectrum can be filtered prior to multiplexing the individual WDM channels without strongly increasing the BER. The bandwidth is then $2/T$. In contrast, the overlapping sinc pulses have a rectangular spectrum that is fully confined to the Nyquist bandwidth $1/T$.

In terms of spectral efficiency, the advantage of pulse shaping becomes immediately apparent when looking at Fig. B.4(c). For the NRZ case, the nearest neighbors need to have a much larger spacing than for the case with sinc-shaping. Or alternatively, for the same frequency grid, the symbol rate of each WDM channel could be increased. Both of these measures increase the SE of the WDM system. A discussion of the options on how to implement pulse shaping can be found in [188].

Single-mode optical fibers are weakly guiding waveguides with circular symmetry that are operated at frequencies below the cut-off frequency of higher-order transverse modes. The lowest-order transverse mode is two-fold degenerate that means there are two orthogonal states of polarization with the same propagation constant. The orthogonality of the polarization states can be used to double the data rate through polarization-division multiplexing (PDM) [189]. Due to imperfections of fiber as well as stress and bending, the circular symmetry of the fiber is broken leading to slightly different propagation constants for the orthogonally polarized modes, i.e., birefringence [117]. Birefringence is randomly changing along the fiber and leads to coupling of the two polarized modes. However, for a given piece of fiber, the polarization changes are unitary in the absence of polarization dependent loss and can be compensated [190], [191].

B.3 Fiber transmission

For transmitting the data over long distances, standard single mode fibers (SSMF) according to ITU recommendation G.652 [192] are used. The two most important characteristics for fiber transmission are loss and chromatic dispersion.

B.3.1 Losses

During propagation along z-direction in a SSMF, losses occur due to scattering and absorption. Introducing the attenuation coefficient α, the average power P of the optical field changes according to Beer's law,

$$\frac{\mathrm{d}P}{\mathrm{d}z} = -\alpha P. \tag{B.8}$$

The output power P_{out} as a function of fiber length L and input power P_{in} can then be expressed as

$$P_{\text{out}} = P_{\text{in}} \exp(-\alpha L). \tag{B.9}$$

The loss coefficient is generally expressed in units of dB/km using [117]

$$\alpha_{\text{dB}} = -\frac{10}{L} \log_{10}\left(\frac{P_{\text{out}}}{P_{\text{in}}}\right) \approx 4.343\alpha. \tag{B.10}$$

For the SSMF, the loss coefficient corresponds to 0.2 dB/km. This amounts to a loss of 15 dB for typical transmission spans of 75 km.

B.3.2 Chromatic dispersion

Chromatic dispersion means that the effective group index of the fiber is frequency dependent. In other words, different spectral components of the data signal travel at a different speed and therefore arrive with a relative delay at the receiver. This effectively deforms the temporal signal shape and leads to inter-symbol interference.

For a simple mathematical treatment, consider a plane wave with carrier frequency $f_c = 2\pi/\omega_c$. Pulse propagation in the z-direction through the dispersive fiber is most easily described in the frequency-domain by multiplication with the frequency dependent propagator $\exp(-\mathrm{j}\beta(\omega)z)$. The propagation constant $\beta(\omega) = \omega n_e(\omega)/c_0$ is related to the effective index $n_e(\omega)$ of the transversal mode of the waveguide, c_0 is the speed of light in vacuum. For signals with a narrow-band optical spectrum it is useful to expand $\beta(\omega)$ in a Taylor series about the carrier frequency ω_c

$$\beta(\omega) = \beta_c^{(0)} + (\omega - \omega_c)\beta_c^{(1)} + \frac{(\omega - \omega_c)^2}{2!}\beta_c^{(2)} + \ldots, \quad \beta_c^{(i)} = \left.\frac{\partial^i \beta}{\partial \omega^i}\right|_{\omega=\omega_c} . \tag{B.11}$$

The first Taylor coefficient is related to the phase velocity v_p

$$\beta_c^{(0)} = \omega_c \frac{n_e(\omega_c)}{c_0} = \frac{\omega_c}{v_p}. \tag{B.12}$$

The second Taylor coefficient represents the inverse of the effective group velocity v_{eg}

$$\begin{aligned} \beta_c^{(1)} &= \frac{1}{c_0} \left.\frac{\partial \omega n_e(\omega)}{\partial \omega}\right|_{\omega=\omega_c} = \frac{1}{c_0}\left.\left(n_e(\omega) + \frac{\partial n_e(\omega)}{\partial \omega}\right)\right|_{\omega=\omega_c} \\ &= \frac{n_{eg}(\omega_c)}{c_0} = \frac{1}{v_{eg}(\omega_c)}. \end{aligned} \tag{B.13}$$

The third (as well as higher order) Taylor coefficient quantifies the dispersion, i.e., the different arrival times t_{eg} of the different spectral components of the optical signal. As discussed in Section A.1, these coefficients can also be used to describe the resonance frequencies of a resonator that is built using a dispersive waveguide.

In most cases of practical interest, the second-order dispersion term $\beta_c^{(2)}$ is dominating[17]. The difference of the arrival times Δt_{eg} of two spectral components at ω_c and $\omega_c + \Delta\omega_c$ is given by

$$\Delta t_{eg} = \left.\frac{\mathrm{d}t_{eg}(\omega)}{\mathrm{d}\omega}\right|_{\omega=\omega_c} \times \Delta\omega_c = \beta_c^{(2)} \Delta\omega_c z\,. \tag{B.14}$$

It is common to define a dispersion parameter C_λ that relates Δt_{eg} of two spectral components to their wavelength separation $\Delta\lambda_c$

$$\frac{\Delta t_{eg}}{z} = C_\lambda \Delta\lambda_c\,, \tag{B.15}$$

which can be expressed in terms of the $\beta_c^{(2)}$

$$C_\lambda = \frac{1}{z}\frac{\Delta t_{eg}}{\Delta\lambda_c} = \frac{1}{z}\frac{\Delta t_{eg}}{\Delta\omega}\frac{\Delta\omega}{\Delta\lambda_c} = \beta_c^{(2)}\left(-\frac{2\pi c}{\lambda_c^2}\right). \tag{B.16}$$

The dispersion parameter C_λ of a SSMF is 17 $\mathrm{ps/nm/km}$, where the group delay difference is given in ps, the propagation length in km and the spectral separation in nm.

B.4 Receiver

To receive and to process the data, the optical signal needs to be converted back into the electrical domain. Photodetectors are used for this purpose. A distinction is made between direct detection and coherent detection. In the former case, the optical signal is mixed with the simultaneously transmitted carrier on the photodetector – square-law detection – thus generating a photocurrent proportional to the power of the optical signal. In the latter case the optical signal is mixed with a locally provided optical carrier (local oscillator). For simultaneous detection of $I(t)$ and $Q(t)$, a second 90° phase shifted version of the same local oscillator is required. In this way the full amplitude and phase information of the optical field can be retrieved. The following paragraphs introduce the receiver implementations that were used

[17] Higher-order dispersion coefficients are mostly relevant at frequencies, where the second-order term is zero. For the SSMF this is near 230 THz, corresponding to a wavelength of 1300 nm. This is far away from the commonly used frequency range of DWDM systems, which are centered at 193.1 THz and span approximately 10 THz.

for the experiments described in Chapters 3 to 5. Finally, the signal quality metrics used to evaluate the system performance will be established.

B.4.1 Direct detection

Direct detection, also referred to as square-law detection, can be used for intensity modulated signals as for example on-off keying (OOK), where the information is encoded in the transmitted optical power. A photodiode generates an output current i that is proportional to the incident external optical power P_e,

$$i = SP_e, \tag{B.17}$$

with the proportionality constant S, commonly referred to as sensitivity or responsivity of the photodiode. The photo-detection process is based on a semiconductor having a bandgap smaller than the photon energy $\hbar\omega$, where ω is the angular frequency of the optical field and $\hbar = h/2\pi$ is the reduced Planck constant. Absorbed photons generate electron-hole pairs, which are separated by an electric field applied across the semiconductor, thus generating the photocurrent i. Physically, the photodiode responsivity is connected to the quantum efficiency η of the photodiode, i.e., the ratio of the rate of electron generation and the rate of incident photons:

$$S = \frac{\eta e}{\hbar\omega}, \tag{B.18}$$

where e is the elementary charge.

The photodiodes used for detection of high-speed data signals typically utilize a pin structure. That means that an intrinsic semiconductor is sandwiched between a p and an n-doped semiconductor. The semiconductor chosen for the intrinsic region typically has a smaller bandgap than the surrounding doped semiconductors such that the absorption takes place only in the intrinsic region. A reverse bias across the pin structure separates the electron-hole pairs. Photodiodes with large sensitivity of 0.5 A/W, large bandwidth of 100 GHz and low dark current of typically 5 nA are commercially available[18].

[18] https://www.finisar.com/optical-components/xpdv412xr

B.4.2 Coherent detection

Recall that for a high spectral efficiency we want to transmit a real signal $E_r(t) = I(t)\cos(\omega_c t) - Q(t)\sin(\omega_c t)$ according to Eq. (B.2). Simple square-law detection as described above would only lead to a photocurrent $i \propto I^2(t) + Q^2(t)$. It is hence not possible to distinguish between $I(t)$ and $Q(t)$ at the receiver. Note that mathematically the square of E_r would contain terms with the frequency $2\omega_c$. However, these terms are removed as the photodetector cannot emit at this frequency.

It is instructive to take a look at the spectrum $\tilde{E}_r(f)$ of the data signal, which can be obtained by Fourier transformation. To this end, Eq. (B.2) is rewritten using exponential functions,

$$\begin{aligned} E_r(t) &= \frac{1}{2} I(t)\left(\mathrm{e}^{\mathrm{j}\omega_c t} + \mathrm{e}^{-\mathrm{j}\omega_c t}\right) + \frac{\mathrm{j}}{2} Q(t)\left(\mathrm{e}^{\mathrm{j}\omega_c t} - \mathrm{e}^{-\mathrm{j}\omega_c t}\right), \\ \tilde{E}_r(f) &= \frac{1}{2}\left[\tilde{I}(f - f_c) + \tilde{I}(f + f_c) + \mathrm{j}\tilde{Q}(f - f_c) - \mathrm{j}\tilde{Q}(f + f_c)\right]. \end{aligned} \tag{B.19}$$

It can be seen that the spectra corresponding to the Fourier transformations of the independent data streams $I(t)$ and $Q(t)$ are symmetrically located around the carrier frequency f_c and are thus overlapping. It is hence not possible to separate the two signals by means of a spectral filter, however, due to the orthogonality between I and Q, they can still be detected separately as explained in the following.

Consider now a different receiver structure where the transmitted signal is multiplied with $\cos(\omega_c t)$ in one case and with $\sin(\omega_c t)$ in the other case. This leads to two photocurrents $i_{\mathrm{I}}(t)$ and $i_{\mathrm{Q}}(t)$ with the following proportionalities

$$\begin{aligned} i_{\mathrm{I}}(t) &\propto \left[I(t)\cos^2(\omega_c t) - Q(t)\sin(\omega_c t)\cos(\omega_c t)\right]^2 = \frac{1}{4} I^2(t), \\ i_{\mathrm{Q}}(t) &\propto \left[I(t)\cos(\omega_c t)\sin(\omega_c t) - Q(t)\sin^2(\omega_c t)\right]^2 = \frac{1}{4} Q^2(t). \end{aligned} \tag{B.20}$$

This technique makes it possible to distinguish between the in-phase and quadrature signal and is commonly referred to as IQ detection.

For this detection method the receiver's local oscillator (LO) needs to be phase-locked to the carrier. This situation is then also referred to as homodyne reception. From a practical point of view it is quite challenging to phase-lock two remotely located lasers. Hence, the carrier and the LO are typically free-running lasers that oscillate at different frequencies. This so-called heterodyne detection establishes an intermediate step in the detection process. The data signal is down-mixed to an intermediate frequency corresponding to the frequency difference between carrier and LO, and the required phase lock is subsequently obtained at the intermediate frequency, where it is easier to achieve. Naturally, this requires a low phase noise of both the carrier and the LO, as discussed in Section 2.2.2.

90° optical hybrid with balanced photodetectors

In analogy to the IQ modulator described in Section B.1, where two independent amplitude modulators are combined with a relative 90° phase shift, a coherent receiver independently detects the I and the Q component of the incoming signal. The basic structure of such a receiver is depicted in Fig. B.5. It consists of a 90° optical hybrid and a pair of balanced photodetectors. The incoming signal is mixed with an LO on one of the balanced photodetectors and with a 90° phase shifted version of the LO on the other balanced photodetector.

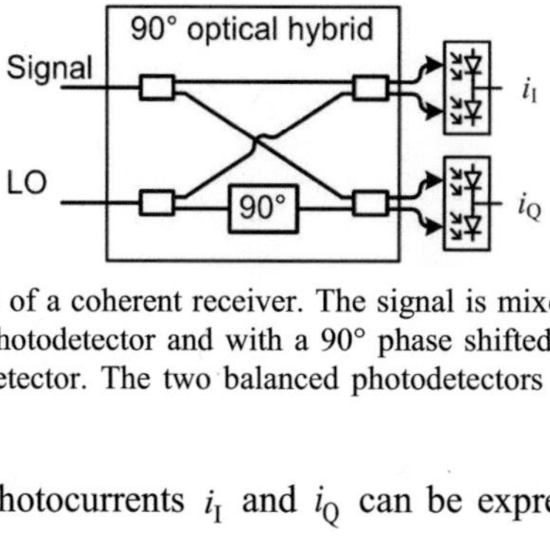

Fig. B.5: Basic elements of a coherent receiver. The signal is mixed with a local oscillator (LO) on one balanced photodetector and with a 90° phase shifted version of the LO on a second balanced photodetector. The two balanced photodetectors generate the currents i_I and i_Q.

The two generated photocurrents i_I and i_Q can be expressed in terms of the electric fields

$$\underline{E}_S = A_S(t)\exp\left[j2\pi f_S t + j\varphi_S(t)\right] \tag{B.21}$$

and

$$\underline{E}_{LO} = A_{LO}(t)\exp\left[j2\pi f_{LO} t + j\varphi_{LO}(t)\right] \tag{B.22}$$

of the signal and the LO, respectively. The photocurrents oscillate at the intermediate frequency $f_{IF} = f_S - f_{LO}$:

$$\begin{aligned} i_{\mathrm{I}}(t) &= 4S\left|A_{\mathrm{S}}(t)A_{\mathrm{LO}}(t)\right|\cos\left(2\pi f_{\mathrm{IF}}t+\varphi_{\mathrm{S}}(t)-\varphi_{\mathrm{LO}}(t)\right) \\ i_{\mathrm{Q}}(t) &= 4S\left|A_{\mathrm{S}}(t)A_{\mathrm{LO}}(t)\right|\sin\left(2\pi f_{\mathrm{IF}}t+\varphi_{\mathrm{S}}(t)-\varphi_{\mathrm{LO}}(t)\right) \end{aligned} \tag{B.23}$$

For a DWDM system, the choice of f_{IF} is very important. It is typically chosen to be as small as possible for the data channel that is to be received. That means, the LO frequency lies within the bandwidth of the signal, a situation that is refered to as intradyne reception [193]. Consider the case of a DWDM system with three closely spaced channels at the center frequencies $f_{\mathrm{S},i}$ as shown in Fig. B.6(a). All channels have the bandwidth B. In order to detect the center channel, $f_{\mathrm{LO}} \approx f_{\mathrm{S},2}$ is chosen. Given that the photodetectors have sufficiently large bandwidth, the spectra of the photocurrents i_{I} and i_{Q} are schematically depicted in Fig. B.6(b).

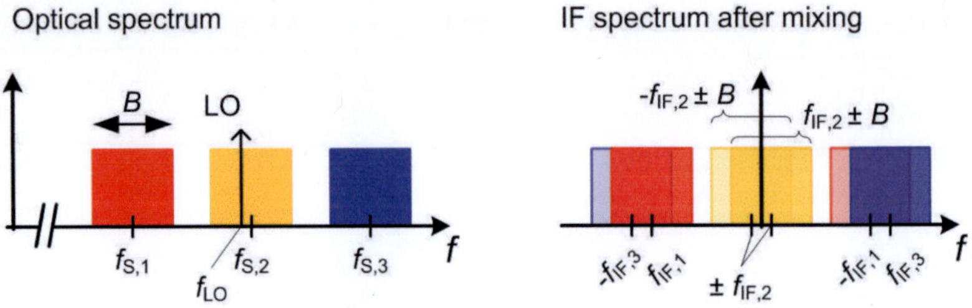

Fig. B.6: Spectra before and after mixing process. (a) Optical spectrum of a DWDM signal with three channels centered at $f_{\mathrm{c},1} \dots f_{\mathrm{c},3}$ and each spanning a bandwidth B. The frequency of the local oscillator f_{LO} is chosen for intradyne detection of the center channel. (b) IF spectrum after the mixing process. Upper and lower sidebands with bandwidth B appear symmetrically at $\pm f_{\mathrm{IF},i} = f_{\mathrm{S},i} - f_{\mathrm{LO}}$. If f_{LO} is chosen close enough to $f_{\mathrm{S},2}$, the center channel can be isolated and demodulated using IQ-detection.

It can be seen that there are spectral components centered at $\pm f_{\mathrm{IF},i}$, each having the bandwidth B. If the local oscillator frequency was chosen close enough to the center frequency of the desired DWDM channel, it is possible to use a low-pass filter to isolate the desired data channel, which can then be demodulated using IQ detection. This requires an additional local oscillator that is phase-locked to $f_{\mathrm{IF},2}$ [193]. Note that also large values for f_{IF} can be chosen if it is possible to filter the desired channel with an optical bandpass filter before entering the 90° hybrid. However, for narrow guard bands, there are no optical filters available with steep-enough edges.

In common coherent optical receivers, the two currents i_{I} and i_{Q} are sampled using fast analog-to-digital converters (ADC) and then processed in the digital domain to retrieve the data [194]. This digital signal processing (DSP) includes also the low-pass filtering and IQ detection. As the full optical field information is available at the receiver, digital compensation of transmission impairments is possible [195]. For the experiments discussed in this thesis, DSP comprises chromatic dispersion compensation, polarization demultiplexing, brick-wall filtering, estimation of f_{IF}, phase estimation and locking, equalization of the system's frequency response, symbol detection and demapping. To this end we use a commercially available coherent receiver, the Keysight N4391A optical modulation analyzer[19]. For this device, high-speed real-time oscilloscopes are used to capture data sequences containing up to five million samples. These are stored and processed offline using the associated software[20].

B.4.3 Signal quality metrics

The most fundamental and relevant signal quality metric is the bit-error-ratio (BER). Irrespective of the type of signal impairment, a low BER represents the most convincing argument to demonstrate that a system is working. Depending on the application, the BER requirements are between 10^{-9} and 10^{-15}. In many cases such low BER cannot be achieved easily, however, there are forward error correction (FEC) techniques [138], [139], [179], [196] that can significantly reduce the BER. This relaxes the system performance requirements, albeit at the expense of transmitting redundant data. It further requires an additional processing effort that adds to system complexity, cost, power consumption and latency. Still the benefits outweigh the drawbacks of using FEC.

For each FEC technique there is a certain threshold BER value that marks the limit that can still be corrected to a post-FEC BER of 10^{-15}. The BER threshold for an FEC with 7 % overhead is 4.5×10^{-3} [179], while a more advanced code works with 20 % overhead and exhibits a BER threshold of 1.5×10^{-2} [139]. Further information on FEC codes and their development history can be found in [138], [196]. Throughout this thesis we compare our measured or estimated BER values to these BER threshold values and take the associated overhead into account in order to distinguish between line rate, i.e.,

---

[19] http://www.keysight.com/en/pd-1593862-pn-N4391A/optical-modulation-analyzer?cc=DE&lc=ger

[20] http://www.keysight.com/en/pd-2405170/stand-alone-software?nid=-33208.1086878.00&cc=DE&lc=ger

the rate at which the transmitter sends the bits, and net data rate, i.e., the rate of information bits that are being transmitted.

Reliably measuring BER is, however, intricate and can be very time consuming, particularly if the BER is low. The transmitted data sequence needs to be known at the receiver to be able to compare the transmitted and the received bits. In order to have a reliable estimate of the BER that should not exceed a maximum value BER_{max} relative to the actual BER, a large number of errors need to be measured. If the measured value should lie in a confidence interval $\left(\mathrm{BER}_{\max} - \mathrm{BER}\right)/\mathrm{BER} = 100\%$, with a probability of 99 %, then at least 13 errors need to be found [197], [198]. This limits especially coherent data transmission experiments with offline processing, where only a limited number of bits can be compared, depending on data rate, ADC sampling rate, and memory size.

If direct BER measurement is not possible, other signal quality metrics might be used to evaluate the system performance. Under certain assumptions, a relation between these metrics and the BER can be established. For this work we use eye diagrams and the Q-factor metric for the OOK measurements, and error-vector magnitude (EVM) for measurements using QPSK and 16QAM. A brief introduction into these metrics and the conversion to BER is given in the following paragraphs.

For an OOK signal using direct-detection, a convenient quality metric is to look at the eye diagram. An exemplary one, taken from [198], is depicted in Fig. B.7. The eye diagram exhibits two horizontal rails that are vertically offset. The lower rail corresponds to the signal level representing a logical '0', while the upper rail corresponds to the signal level that represents a logical '1'. Transitions between the two levels are also shown. Centered in-between the transitions is the optimum sampling point for the decision circuit. The blue and red curves on the right hand side of the eye diagram correspond to histograms of the two signal levels. They represent the probability density function $w_{0,1}(u)$ for measuring a voltage u at the sampling point. In many cases of practical interest, $w_{0,1}(u)$ can be approximated by a Gaussian distribution with the mean values $u_{0,1}$ and variances $\sigma_{0,1}^2$ for the two levels, respectively [198]

$$w_{0,1}(u) = \frac{1}{\sqrt{2\pi\sigma_{0,1}^2}} \exp\left[\frac{\left(u - u_{0,1}\right)^2}{2\sigma_{0,1}^2}\right]. \tag{B.24}$$

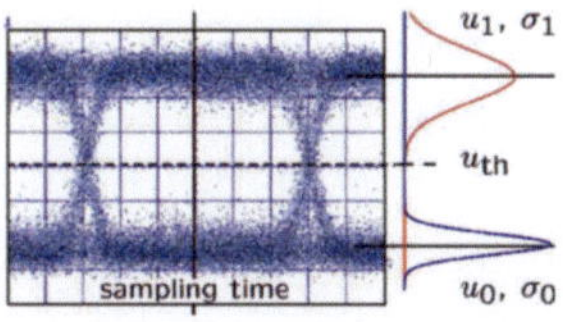

Fig. B.7: Eye diagram taken from [198]. The eye diagram shows a lower rail with average voltage u_0 and standard deviation σ_0 corresponding to the signal level that represents a logical '0'. Likewise, the upper rail with voltage u_1 and standard deviation σ_1 corresponds to the signal level representing a logical '1'. The optimum sampling time is centered between the transitions amongst the '0' and '1' level. The voltage u_{th} represents the optimum threshold voltage for the decision circuit. The red and blue histograms on the right hand side represent the probability density function $w_{0,1}$ of the two signal levels.

The Q-factor metric is a measure of the eye opening. It is defined as

$$Q = \frac{u_1 - u_0}{\sigma_1 + \sigma_0}. \tag{B.25}$$

Assuming that the ratio of the occurrence probabilities of transmitted ones $p(1t)$ and zeros $p(0t)$ is equal to the ratio of the associated standard deviations and an optimum decision threshold u_{th} along with the Gaussian probability distribution according to Eq. (B.24), the Q-factor can be related to the BER [198]

$$\text{BER}=\frac{1}{2}\text{erfc}\left(\frac{Q}{\sqrt{2}}\right), u_{th} = \frac{u_1\sigma_0 + u_0\sigma_1}{\sigma_0 + \sigma_1} \text{ for } \frac{p(1t)}{p(0t)} = \frac{\sigma_1}{\sigma_0}. \tag{B.26}$$

The complementary error-function $\text{erfc}(z)$ is defined by

$$\text{erfc}(z) = \frac{2}{\sqrt{\pi}}\int_z^\infty \exp\left[-(z')^2\right] \mathrm{d}z'. \tag{B.27}$$

To characterize the signal quality of higher-order modulation formats, it is insightful to look at the measured constellation diagram, which provides additional information about the sources of signal degradation. If the cloud of measured symbols is elongated in azimuthal direction, this hints to a dominant phase noise, while an elongation in radial direction stems from intensity noise. For the case of dominating additive white Gaussian noise, the overlay of many measured constellation points forms circular clouds with a size in proportion to the optical signal-to-noise power ratio (OSNR).

For quantification purposes, it is common to use the error-vector magnitude (EVM). The EVM is defined as the root mean square average of the error vector of all received symbols, normalized either to the maximum field $|\underline{E}_{\mathrm{t,m}}|$ in the constellation, or to the average field, which is obtained by summing the powers of all M possible symbols. Fig. B.8 shows the definition of the error vector $\underline{E}_{\mathrm{err}} = \underline{E}_{\mathrm{r}} - \underline{E}_{\mathrm{t}}$, i.e., the vector from the transmitted symbol to the received point in the IQ-plane.

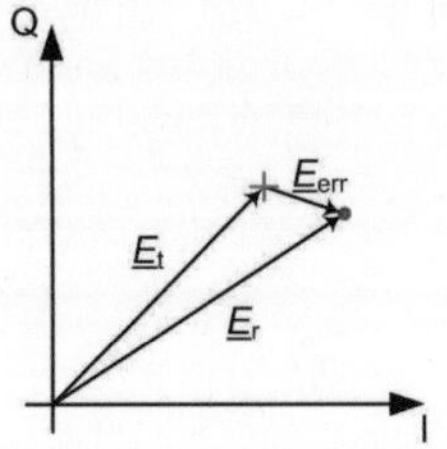

Fig. B.8: Definition of the error vector $\underline{E}_{\mathrm{err}}$ in the IQ-plane. The green cross denotes the transmitted symbol with vector $\underline{E}_{\mathrm{t}}$ and the red dot shows the received vector $\underline{E}_{\mathrm{r}}$.

The EVM is then given as

$$\mathrm{EVM_m} = \frac{\sqrt{\frac{1}{N}\sum_{i=1}^{N}\left|\underline{E}_{\mathrm{err},i}\right|^2}}{\left|\underline{E}_{\mathrm{t,m}}\right|}, \tag{B.28}$$

where the subscript ‘m’ stands for the normalization to the maximum field in the constellation. This normalization is used throughout this thesis, sometimes also without the subscript. According to [114], [115], the BER can be approximated from the $\mathrm{EVM_m}$ as

$$\mathrm{BER} \approx \frac{\left(1 - M^{-\frac{1}{2}}\right)}{\frac{1}{2}\log_2 M}\,\mathrm{erfc}\left[\sqrt{\frac{3}{2(M-1)}\frac{1}{\left(k\mathrm{EVM_m}\right)^2}}\right], \tag{B.29}$$

for the case of quadratic constellations, additive white Gaussian noise as the main source of signal degradation and data-aided reception. In Eq. (B.29), the factor k accounts for the different normalizations of the EVM and is $k = 1$ for QPSK and $k = \sqrt{9/5}$ for 16QAM. M is the constellation size, or in other words, $\log_2 M$ is the number of bits encoded into each symbol.

Appendix C Noise in laser oscillators

Consider a laser oscillator with central frequency f_s, amplitude $\rho(t)$ and phase $\varphi(t)$. If the electric field $S(t) = \rho(t)\cos\left[2\pi f_s t + \varphi(t)\right]$ of this laser is mixed with two ideal signals that are proportional to $\cos 2\pi f_s t$ and $\sin 2\pi f_s t$, then a complex phasor $S = \rho\cos\varphi + \mathrm{j}\rho\sin\varphi$ can be obtained. Amplitude ρ and phase φ are random variables which depend statistically on time t. For incoherent light, i.e., for a light emitting diode (LED) or for a laser far below threshold, the phasor tip executes a two-dimensional random walk at a rate inversely proportional to the bandwidth of the radiation. Observing S over a longer period, the statistical location of S follows a joint probability density function (pdf) $w_{\rho\varphi}(\rho,\varphi)$ in the form of a circular Gaussian "mole-hill" [199] centered at the origin, see Fig. C.1(a).

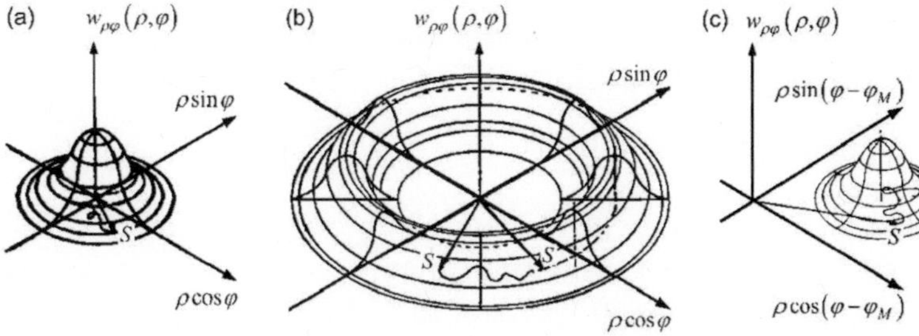

Fig. C.1: Statistics of laser oscillators using the joint probability density function (pdf) $w_{\rho\varphi}(\rho,\varphi)$ of the electric field amplitude ρ and phase φ plotted over the real part $\rho\cos\varphi$ and imaginary part $\rho\sin\varphi$ of the electric field. (a) Far below threshold, the laser shows the same behavior as an LED. Amplitude and phase follow a diffusion process and the joint pdf resembles a Gaussian in each direction ("mole-hill"). The expectation value of the amplitude is $\bar{\rho} = 0$ and the phase is equally distributed within $0 \le \varphi < 2\pi$. (b) Far above threshold, the joint pdf resembles a ring wall ("mole-run"). The oscillator shows a tendency to regulate the amplitude to a constant value, while the phase is still subject to a random drift. (c) Joint pdf of an injection-locked laser, where the phase φ is locked to the phase φ_M of a master laser causing a restoring force for the phase as soon as it deviates from φ_M. Figures are modified from [C3].

In contrast to that, the joint pdf for coherent light, i.e., laser oscillation far above threshold, will be a circular "mole-run" centered at the origin, but with vanishingly small probability at the origin, see Fig. C.1(b). The radial cross-section is generally Gaussian. The amplitude stabilizes at a constant expectation value $\bar{\rho}$ due to the regenerative property of the oscillator. This can be understood as follows: If the amplitude ρ increases (decreases), then the inversion of the laser's gain medium decreases (increases) forcing ρ back to $\bar{\rho}$. Conversely, for the phase there is no intrinsic restoring force that is proportional to the departure of the phasor from a certain phase reference. Hence the equal distribution of the phase within $0 \leq \varphi < 2\pi$.

It is possible to provide such a restoring force for the phase by injecting light from an external "master" laser with low phase noise, if the frequency f_M of the master laser is close to the free-running laser frequency f_s. The linear superposition of both fields inside the laser cavity generates an oscillation of the carrier density at $|f_M - f_s|$. This periodic change of the gain mixes with the laser's native oscillation frequency and therefore pulls the emission frequency from f_s to f_M. The phase difference $\Delta\varphi = \varphi - \varphi_M$ between master and slave remains constant once they are locked, hence the joint pdf of the injection-locked laser resembles a "mole-hill" at a constant offset from the origin, see Fig. C.1(c). Phase locking is only possible if the condition $|\Delta\varphi| < \pi/2$ is maintained, which requires sufficiently small differences $|f_M - f_s|$ of the initial native frequencies [200].

C.1 Linewidth

Random fluctuations of the phase cause a broadening of the spectral width of the laser line. This can be understood from the fact that the instantaneous frequency f_{inst} is the time-derivative of the phase: $2\pi f_{\text{inst}}(t) = \mathrm{d}\varphi/\mathrm{d}t$. There are two main mechanisms that cause phase fluctuations in semiconductor lasers. Both are connected to spontaneous emission. On one hand the spontaneous emission directly causes a random change of the phase (and intensity) of the lasing field. On the other hand, there will also be a delayed phase change due to amplitude-phase coupling in the laser [201]. This can be understood as follows: The spontaneous emission causes fluctuations of the laser's amplitude. This will in turn change the carrier density as the laser thrives to restore the steady-state amplitude. As the carrier density affects the refractive index of the active medium, there will be also be a phase change. The linewidth broadening caused by this amplitude-phase coupling is quantified by the so-called Henry factor (or linewidth enhancement factor) α_H and amounts to $1 + \alpha_H^2$ [201]. The lineshape is found to be Lorentzian with linewidth Δf that depends on the rate R_{sp} of spontaneous emission into the lasing mode, and on the mean photon number $\overline{N_p}$ [117]

$$\Delta f = R_{\text{sp}} \frac{\left(1+\alpha_H^2\right)}{4\pi \overline{N_p}}. \tag{C.1}$$

C.2 Relative intensity noise

Semiconductor lasers typically feature classical intensity noise, which can be modeled as a small perturbation $\delta P(t)$ to its mean output power $\overline{P}$. These fluctuations are typically caused by amplified spontaneous emission inside the laser cavity. It is common to quantify the relative intensity noise (RIN) as the ratio of the variance of the laser power fluctuations $\sigma_P^2 = \overline{\delta P^2}$ and the square of the mean output power $\overline{P}^2$. RIN has a one-sided power spectrum $\text{RIN}(f)$

$$\text{RIN} = \int_0^{\infty} \text{RIN}(f)\,\mathrm{d}f = \frac{\overline{\delta P^2}}{\overline{P}^2}. \tag{C.2}$$

The RIN can be measured by detecting the light with a photodiode and relating the mean current with the variance of its fluctuations. As discussed in Section B.4.1, a photodiode converts optical power P to an electrical current i with the proportionality constant S. The generated photocurrent will therefore comprise a mean part $\bar{i} = S\overline{P}$ and fluctuations with a variance $\sigma_i^2 = S\sigma_P^2 = S\overline{\delta P^2}$. In addition to the fluctuations of the optical signal, shot noise and thermal noise also contribute to the total variance of the photocurrent.

We assume that the noise power spectral density S_N of the light is approximately constant over a bandwidth B_N, leading to the noise power $P_N = S_N B_N$. The (noisy) carrier is detected with a receiver having the bandwidth $B \ll B_N$. There are several contributions to the variance of the photocurrent [117], [202]:

1) Thermal noise, which depends on the temperature T_r and the load resistance R_L of the photodiode located in the front end of an optical receiver. The Boltzmann constant is denoted by k_B:

$$\overline{\left|i_{\text{th}}\right|^2} = \frac{4k_B T_r B}{R_L} \tag{C.3}$$

2) Shot noise generated by the carrier as well as the noise[21]:

$$\overline{|i_{\text{shot}}|^2} = 2eS(P + P_N)B \tag{C.4}$$

3) Mixing of the carrier and the copolarized part of the noise, i.e., only half of S_N contributes for the case of unpolarized noise:

$$\overline{|i_{\text{c-N}}|^2} = 4S^2P(S_N/2)B \tag{C.5}$$

4) Mixing of the noise with itself:

$$\overline{|i_{\text{N-N}}|^2} = S^2S_N^2\left(B_N - \frac{B}{2}\right)B \tag{C.6}$$

For sufficiently large P and P_N, the contribution from $\overline{|i_{\text{c-N}}|^2}$ is dominating [117]. In this case RIN can be directly related to the optical carrier-to-noise-density power ratio (OCN_0R) that is defined as the ratio of P and S_N.

$$\text{RIN} = \frac{\overline{\delta P^2}}{\overline{P}^2} = \frac{\overline{|i_{\text{c-N}}|^2}}{\overline{i}^2} = \frac{2S^2PS_NB}{S^2P^2} = 2B\frac{S_N}{P} = \frac{2B}{\text{OCN}_0\text{R}} \tag{C.7}$$

The electrical signal-to-noise power ratio (SNR) is then directly related to the relative intensity noise

$$\text{SNR} = \frac{\overline{i}^2}{\sigma_i^2} = \frac{\overline{P}^2}{\overline{\delta P^2}} = \frac{1}{\text{RIN}} \tag{C.8}$$

[21] Shot noise requires that the photon probability distribution is a Poisson distribution, which is true for an ideal coherent carrier, but not for the spontaneous emission noise, where the photon probability follows a Bose-Einstein distribution. However, if the bandwidth B_N of the noise is much larger than the receiver bandwidth B then only few photons occupy each longitudinal mode of the fiber. In this case the photon probability distribution can be approximated by a Poisson distribution [202].

The RIN can be measured using a photodiode, a bias-T and an electrical spectrum analyzer (ESA) [203]. The photodiode detects the light, and the generated photocurrent I_{PD} can be measured at the DC-port of the bias-T. At the same time the ESA measures the power spectral density (PSD) of the fluctuations of the photocurrent. Assuming a 50 Ω termination of the photodiode we find

$$\mathrm{RIN}(f)=\frac{\mathrm{PSD}(f)}{I_{\mathrm{PD}}^{2}\cdot 50\Omega}. \tag{C.9}$$

Appendix D Bibliography

[1] Cisco Systems Inc., "The Zettabyte Era—Trends and Analysis," 2015.

[2] Cisco Systems Inc., "Cisco Global Cloud Index: Forecast and Methodology, 2013–2018," 2014.

[3] Alcatel Lucent Strategic White Paper, "METRO NETWORK TRAFFIC GROWTH: AN ARCHITECTURE IMPACT STUDY," 2013.

[4] W. Kaiser, T. Kupfer, A. Herzberger, C. R. S. Fludger, J. C. Geyer, T. Duthel, and C. Schulien, "Integrated circuits for coherent transceivers for 100G and beyond," *Opt. Fiber Technol.*, vol. 17, no. 5, pp. 456–463, 2011.

[5] J. Geyer, C. R. Doerr, M. Aydinlik, N. Nadarajah, A. Caballero, C. Rasmussen, and B. Mikkelsen, "Practical Implementation of Higher Order Modulation Beyond 16-QAM," in *Optical Fiber Communication Conference*, 2015, p. Th1B.1.

[6] L. Kull, "Challenges in implementing high-speed, low-power ADCs," in *Optical Fiber Communication Conference*, 2015, p. Th1B.2.

[7] R.-J. Essiambre, G. Kramer, P. J. Winzer, G. J. Foschini, and B. Goebel, "Capacity limits of optical fiber networks," *J. Lightwave Technol.*, vol. 28, no. 4, pp. 662–701, 2010.

[8] K. Roberts, M. O'Sullivan, K. T. Wu, H. Sun, A. Awadalla, D. J. Krause, and C. Laperle, "Performance of dual-polarization QPSK for optical transport systems," *J. Lightwave Technol.*, vol. 27, no. 16, pp. 3546–3559, 2009.

[9] K. Roberts, A. Borowiec, and C. Laperle, "Technologies for optical systems beyond 100G," *Opt. Fiber Technol.*, vol. 17, no. 5, pp. 387–394, 2011.

[10] K. Roberts, S. H. Foo, M. Moyer, M. Hubbard, A. Sinclair, J. Gaudette, and C. Laperle, "High Capacity Transport - 100G and Beyond," *J. Lightwave Technol.*, vol. 33, no. 3, pp. 563–578, 2015.

[11] P. Winzer, "Beyond 100G Ethernet," *IEEE Commun. Mag.*, vol. 48, no. 7, pp. 26–30, 2010.

[12] P. J. Winzer, "High-Spectral-Efficiency Optical Modulation Formats," *J. Lightwave Technol.*, vol. 30, no. 24, pp. 3824–3835, 2012.

[13] M. Mitchell, J. McNicol, V. Dangui, H. Sun, K.-T. Wu, Z. Pan, M. Van Leeuwen, J. Rahn, S. Grubb, R. Nagarajan, and D. Welch, "Optical integration and multi-carrier solutions for 100G and beyond," *Opt. Fiber*

Technol., vol. 17, no. 5, pp. 412–420, 2011.

[14] G. Bennett, K. Wu, A. Malik, S. Roy, and A. Awadalla, "A review of high-speed coherent transmission technologies for long-haul DWDM transmission at 100G and beyond," *IEEE Commun. Mag.*, vol. 52, no. 10, pp. 102–110, 2014.

[15] T. Healy, F. C. G. Gunning, A. D. Ellis, and J. D. Bull, "Multi-wavelength source using low drive-voltage amplitude modulators for optical communications," *Opt. Express*, vol. 15, no. 6, pp. 2981–2986, 2007.

[16] P. J. Delfyett, S. Gee, C. Myoung-Taek, H. Izadpanah, L. Wangkuen, S. Ozharar, F. Quinlan, and T. Yilmaz, "Optical frequency combs from semiconductor lasers and applications in ultrawideband signal processing and communications," *J. Lightwave Technol.*, vol. 24, no. 7, pp. 2701–2719, 2006.

[17] N. Alic, E. Myslivets, E. Temprana, B. P.-P. Kuo, and S. Radic, "Nonlinearity Cancellation in Fiber Optic Links Based on Frequency Referenced Carriers," *J. Lightwave Technol.*, vol. 32, no. 15, pp. 2690–2698, 2014.

[18] E. Temprana, E. Myslivets, B. P.-P. Kuo, L. Liu, V. Ataie, N. Alic, and S. Radic, "Overcoming Kerr-induced capacity limit in optical fiber transmission," *Science*, vol. 348, no. 6242, pp. 1445–1448, 2015.

[19] E. Temprana, E. Myslivets, L. Liu, V. Ataie, A. Wiberg, B. P. P. Kuo, N. Alic, and S. Radic, "Two-fold transmission reach enhancement enabled by transmitter-side digital backpropagation and optical frequency comb-derived information carriers," *Opt. Express*, vol. 23, no. 16, pp. 20774–20783, 2015.

[20] ITU-T Recommendation G.694.1, "Spectral grids for WDM applications: DWDM frequency grid." 2012.

[21] S. Gringeri, B. Basch, V. Shukla, R. Egorov, and T. Xia, "Flexible architectures for optical transport nodes and networks," *IEEE Commun. Mag.*, vol. 48, no. 7, pp. 40–50, 2010.

[22] B. Teipen, M. Eiselt, K. Grobe, and J.-P. Elbers, "Adaptive Data Rates for Flexible Transceivers in Optical Networks," *J. Networks*, vol. 7, no. 5, 2012.

[23] X. Zhao, V. Vusirikala, B. Koley, V. Kamalov, and T. Hofmeister, "The prospect of inter-data-center optical networks," *IEEE Commun. Mag.*, vol. 51, no. 9, pp. 32–38, 2013.

[24] A. Lord, P. Wright, and A. Mitra, "Core Networks in the Flexgrid Era," *J.*

Lightwave Technol., vol. 33, no. 5, pp. 1126–1135, 2015.

[25] R. Antosik, “Super-channel architectures for in-service capacity expansion of CWDM/DWDM systems,” in *International Conference on Transparent Optical Networks*, 2003, vol. 2, pp. 84–86.

[26] S. Chandrasekhar and X. Liu, “Advances in Tb/s Superchannels,” in *OPTICAL FIBER TELECOMMUNICATIONS VIB: SYSTEMS AND NETWORKS*, 6th ed., I. Kaminow, T. Li, and A. E. Willner, Eds. Academic Press, 2013, pp. 83–119.

[27] H. Soda, K. Iga, C. Kitahara, and Y. Suematsu, “GaInAsP/InP Surface Emitting Injection Lasers,” *Jpn. J. Appl. Phys.*, vol. 18, no. 12, p. 2329, 1979.

[28] K. Iga, F. Koyama, and S. Kinoshita, “Surface emitting semiconductor lasers,” *IEEE J. Quantum Elect.*, vol. 24, no. 9, pp. 1845–1855, 1988.

[29] A. Syrbu, A. Mircea, A. Mereuta, A. Caliman, C.-A. Berseth, G. Suruceanu, V. Iakovlev, M. Achtenhagen, A. Rudra, and E. Kapon, “1.5-mW single-mode operation of wafer-fused 1550-nm VCSELs,” *IEEE Photonic. Tech. L.*, vol. 16, no. 5, pp. 1230–1232, 2004.

[30] G. Meloni, A. Malacarne, F. Fresi, and L. Poti, “6.27 bit/s/Hz spectral efficiency VCSEL-based coherent communication over 800km of SMF,” in *Optical Fiber Communications Conference and Exhibition (OFC), 2015*, 2015, p. Th2A.30.

[31] R. Nagarajan, M. Kato, J. Pleumeekers, P. Evans, S. Corzine, S. Hurtt, A. Dentai, S. Murthy, M. Missey, R. Muthiah, R. A. Salvatore, C. Joyner, R. Schneider, M. Ziari, F. Kish, and D. Welch, “InP Photonic Integrated Circuits,” *IEEE J. Sel. Top. Quant.*, vol. 16, no. 5, pp. 1113–1125, 2010.

[32] M. Mitchell, J. McNicol, V. Dangui, H. Sun, K.-T. Wu, Z. Pan, M. Van Leeuwen, J. Rahn, S. Grubb, R. Nagarajan, and D. Welch, “Optical integration and multi-carrier solutions for 100G and beyond,” *Opt. Fiber Technol.*, vol. 17, no. 5, pp. 412–420, 2011.

[33] R. Nagarajan, M. Kato, D. Lambert, P. Evans, S. Corzine, V. Lal, J. Rahn, A. Nilsson, M. Fisher, M. Kuntz, J. Pleumeekers, A. Dentai, H.-S. Tsai, D. Krause, H. Sun, K.-T. Wu, M. Ziari, T. Butrie, M. Reffle, M. Mitchell, F. Kish, and D. Welch, “Terabit/s class InP photonic integrated circuits,” *Semicond. Sci. Technol.*, vol. 27, no. 9, p. 94003, 2012.

[34] J. Summers, T. Vallaitis, P. Evans, M. Ziari, P. Studenkov, M. Fisher, J. Sena, A. James, S. Corzine, D. Pavinski, J. O. Yang, M. Missey, D. Gold, D. Lambert, W. Williams, M. Lai, F. Kish, and D. Welch, “40 Channels x 57

Gb/s monolithically integrated InP-based coherent photonic transmitter," *European Conference on Optical Communication (ECOC)*. p. P.2.5, 2014.

[35] C. R. Cole, "100-Gb/s and beyond transceiver technologies," *Opt. Fiber Technol.*, vol. 17, no. 5, pp. 472–479, 2011.

[36] H. Park, A. Fang, S. Kodama, and J. Bowers, "Hybrid silicon evanescent laser fabricated with a silicon waveguide and III-V offset quantum wells," *Opt. Express*, vol. 13, no. 23, pp. 9460–9464, 2005.

[37] A. W. Fang, E. Lively, Y.-H. Kuo, D. Liang, and J. E. Bowers, "A distributed feedback silicon evanescent laser," *Opt. Express*, vol. 16, no. 7, pp. 4413–4419, 2008.

[38] X. Zheng, S. Lin, Y. Luo, J. Yao, G. Li, S. S. Djordjevic, J. H. Lee, H. D. Thacker, I. Shubin, K. Raj, J. E. Cunningham, and A. V. Krishnamoorthy, "Efficient WDM laser sources towards terabyte/s silicon photonic interconnects," *J. Lightwave Technol.*, vol. 31, no. 24. pp. 4142–4154, 2013.

[39] G. Kurczveil, M. J. R. Heck, J. D. Peters, J. M. Garcia, D. Spencer, and J. E. Bowers, "An integrated hybrid silicon multiwavelength AWG laser," *IEEE J. Sel. Top. Quant.*, vol. 17, no. 6, pp. 1521–1527, 2011.

[40] Y. Zhang, S. Yang, X. Zhu, Q. Li, H. Guan, P. Magill, K. Bergman, T. Baehr-Jones, and M. Hochberg, "Quantum dot SOA/silicon external cavity multi-wavelength laser," *Opt. Express*, vol. 23, no. 4, pp. 4666–4671, 2015.

[41] S. Keyvaninia, S. Uvin, M. Tassaert, X. Fu, S. Latkowski, J. Mariën, L. Thomassen, F. Lelarge, G. Duan, P. Verheyen, G. Lepage, J. Van Campenhout, E. Bente, and G. Roelkens, "Narrow-linewidth short-pulse III-V-on-silicon mode-locked lasers based on a linear and ring cavity geometry," *Opt. Express*, vol. 23, no. 3, pp. 3221–3229, 2015.

[42] S. Uvin, S. Keyvaninia, F. Lelarge, G.-H. Duan, B. Kuyken, and G. Roelkens, "Narrow line width frequency comb source based on an injection-locked III–V-on-silicon mode-locked laser," *Opt. Express*, vol. 24, no. 5, pp. 5277–5286, 2016.

[43] J. Müller, J. Hauck, B. Shen, S. Romero-García, E. Islamova, S. S. Azadeh, S. Joshi, N. Chimot, A. Moscoso-Mártir, F. Merget, F. Lelarge, and J. Witzens, "Silicon photonics WDM transmitter with single section semiconductor mode-locked laser," *Adv. Opt. Technol.*, vol. 4, no. 2, pp. 119–145, 2015.

[44] R. Schmogrow, M. Winter, M. Meyer, D. Hillerkuss, S. Wolf, B. Baeuerle, A. Ludwig, B. Nebendahl, S. Ben-Ezra, J. Meyer, M. Dreschmann, M. Huebner, J. Becker, C. Koos, W. Freude, and J. Leuthold, "Real-time Nyquist pulse generation beyond 100 Gbit/s and its relation to OFDM," *Opt. Express*, vol. 20, no. 1, pp. 317–337, 2012.

[45] B. E. Little, J. S. Foresi, G. Steinmeyer, E. R. Thoen, S. T. Chu, H. A. Haus, E. P. Ippen, L. C. Kimerling, and W. Greene, "Ultra-compact Si-SiO2 microring resonator optical channel dropping filters," *IEEE Photonic. Tech. L.*, vol. 10, no. 4, pp. 549–551, 1998.

[46] X. Zheng, I. Shubin, G. Li, T. Pinguet, A. Mekis, J. Yao, H. Thacker, Y. Luo, J. Costa, K. Raj, J. E. Cunningham, and A. V Krishnamoorthy, "A tunable 1x4 silicon CMOS photonic wavelength multiplexer/demultiplexer for dense optical interconnects," *Opt. Express*, vol. 18, no. 5, pp. 5151–5160, 2010.

[47] D. Feng, N.-N. Feng, C.-C. Kung, H. Liang, W. Qian, J. Fong, B. J. Luff, and M. Asghari, "Compact single-chip VMUX/DEMUX on the silicon-on-insulator platform," *Opt. Express*, vol. 19, no. 7, pp. 6125–6130, 2011.

[48] K. Sasaki, F. Ohno, A. Motegi, and T. Baba, "Arrayed waveguide grating of 70×60 μm^2 size based on Si photonic wire waveguides," *Electron. Lett.*, vol. 41, no. 14, pp. 801–802, 2005.

[49] D. W. Kim, A. Barkai, R. Jones, N. Elek, H. Nguyen, and A. Liu, "Silicon-on-insulator eight-channel optical multiplexer based on a cascade of asymmetric Mach-Zehnder interferometers," *Opt. Lett.*, vol. 33, no. 5, pp. 530–532, 2008.

[50] D. Hillerkuss, R. Schmogrow, T. Schellinger, M. Jordan, M. Winter, G. Huber, T. Vallaitis, R. Bonk, P. Kleinow, F. Frey, M. Roeger, S. Koenig, A. Ludwig, A. Marculescu, J. Li, M. Hoh, M. Dreschmann, J. Meyer, B. Ezra S., N. Narkiss, B. Nebendahl, F. Parmigiani, P. Petropoulos, B. Resan, A. Oehler, K. Weingarten, T. Ellermeyer, J. Lutz, M. Moeller, M. Huebner, J. Becker, C. Koos, W. Freude, and J. Leuthold, "26 Tbit s-1 line-rate super-channel transmission utilizing all-optical fast Fourier transform processing," *Nature Photon.*, vol. 5, no. 6, pp. 364–371, 2011.

[51] J. Yu, Z. Dong, J. Zhang, X. Xiao, H.-C. Chien, and N. Chi, "Generation of Coherent and Frequency-Locked Multi-Carriers Using Cascaded Phase Modulators for 10 Tb/s Optical Transmission System," *J. Lightwave Technol.*, vol. 30, no. 4, pp. 458–465, 2012.

[52] D. Hillerkuss, R. Schmogrow, M. Meyer, S. Wolf, M. Jordan, P. Kleinow, N. Lindenmann, P. C. Schindler, A. Melikyan, X. Yang, S. Ben-Ezra, B.

Nebendahl, M. Dreschmann, J. Meyer, F. Parmigiani, P. Petropoulos, B. Resan, A. Oehler, K. Weingarten, L. Altenhain, T. Ellermeyer, M. Moeller, M. Huebner, J. Becker, C. Koos, W. Freude, and J. Leuthold, "Single-Laser 32.5 Tbit/s Nyquist WDM transmission," *J. Opt. Commun. Netw.*, vol. 4, no. 10, pp. 715–723, 2012.

[53] V. Ataie, E. Temprana, L. Liu, E. Myslivets, B. P.-P. Kuo, N. Alic, and S. Radic, "Ultrahigh Count Coherent WDM Channels Transmission Using Optical Parametric Comb Based Frequency Synthesizer," *J. Lightwave Technol.*, vol. 33, no. 3, pp. 694–699, 2015.

[54] H. Ito, H. Yokoyama, S. Murata, and H. Inaba, "Picosecond optical pulse generation from an R. F. modulated AlGaAs D. H. diode laser," *Electron. Lett.*, vol. 15, no. 23, pp. 738–740, 1979.

[55] P. Torphammar and S. T. Eng, "Picosecond pulse generation in semiconductor lasers using resonance oscillation," *Electron. Lett.*, vol. 16, no. 15, pp. 587–589, 1980.

[56] S. Tarucha and K. Otsuka, "Response of semiconductor laser to deep sinusoidal injection current modulation," *IEEE J. Quantum Elect.*, vol. 17, no. 5, pp. 810–816, 1981.

[57] S. Mohrdiek, H. Burkhard, and H. Walter, "Chirp reduction of directly modulated semiconductor lasers at 10 Gb/s by strong CW light injection," *J. Lightwave Technol.*, vol. 12, no. 3, pp. 418–424, 1994.

[58] P. M. Anandarajah, R. Maher, Y. Q. Xu, S. Latkowski, J. O'Carroll, S. G. Murdoch, R. Phelan, J. O'Gorman, and L. P. Barry, "Generation of Coherent Multicarrier Signals by Gain Switching of Discrete Mode Lasers," *IEEE Photon. J.*, vol. 3, no. 1, pp. 112–122, 2011.

[59] R. Zhou, S. Latkowski, J. O'Carroll, R. Phelan, L. P. Barry, and P. Anandarajah, "40nm wavelength tunable gain-switched optical comb source," *Opt. Express*, vol. 19, no. 26, pp. B415--B420, 2011.

[60] G. Yabre, H. de Waardt, H. P. A. den Boom, and G.-D. Khoe, "Noise characteristics of single-mode semiconductor lasers under external light injection," *IEEE J. Quantum Elect.*, vol. 36, no. 3, pp. 385–393, 2000.

[61] D. M. Pataca, P. Gunning, M. L. Rocha, J. K. Lucek, R. Kashyap, K. Smith, D. G. Moodie, R. P. Davey, R. F. Souza, and A. S. Siddiqui, "Gain-switched DFB lasers," *J. Microwaves, Optoelectron. Electromagn. Appl.*, vol. 1, no. 1, pp. 46–63, 1997.

[62] F. Lelarge, B. Dagens, J. Renaudier, R. Brenot, A. Accard, F. van Dijk, D. Make, O. Le Gouezigou, J. Provost, F. Poingt, J. Landreau, O. Drisse, E.

Derouin, B. Rousseau, F. Pommereau, and G.-H. Duan, "Recent Advances on InAs/InP Quantum Dash Based Semiconductor Lasers and Optical Amplifiers Operating at 1.55 μm," *IEEE J. Sel. Top. Quant.*, vol. 13, no. 1, pp. 111–124, 2007.

[63] R. Rosales, K. Merghem, A. Martinez, A. Akrout, J. P. Tourrenc, A. Accard, F. Lelarge, and A. Ramdane, "InAs/InP quantum-dot passively mode-locked lasers for 1.55-μm applications," *IEEE J. Sel. Top. Quant.*, vol. 17, no. 5, pp. 1292–1301, 2011.

[64] J. Renaudier, G.-H. Duan, P. Landais, and P. Gallion, "Phase Correlation and Linewidth Reduction of 40 GHz Self-Pulsation in Distributed Bragg Reflector Semiconductor Lasers," *IEEE J. Quantum Elect.*, vol. 43, no. 2, pp. 147–156, 2007.

[65] R. Rosales, S. G. Murdoch, R. T. Watts, K. Merghem, A. Martinez, F. Lelarge, A. Accard, L. P. Barry, and A. Ramdane, "High performance mode locking characteristics of single section quantum dash lasers," *Opt. Express*, vol. 20, no. 8, pp. 8649–8657, 2012.

[66] P. Del'Haye, A. Schliesser, O. Arcizet, T. Wilken, R. Holzwarth, and T. J. Kippenberg, "Optical frequency comb generation from a monolithic microresonator," *Nature*, vol. 450, pp. 1214–1217, 2007.

[67] P. Del'Haye, T. Herr, E. Gavartin, M. L. Gorodetsky, R. Holzwarth, and T. J. Kippenberg, "Octave Spanning Tunable Frequency Comb from a Microresonator," *Phys. Rev. Lett.*, vol. 107, no. 6, p. 63901, 2011.

[68] T. Herr, K. Hartinger, J. Riemensberger, C. Y. Wang, E. Gavartin, R. Holzwarth, M. L. Gorodetsky, and T. J. Kippenberg, "Universal formation dynamics and noise of Kerr-frequency combs in microresonators," *Nature Photon.*, vol. 6, no. 7, pp. 480–487, 2012.

[69] T. Herr, V. Brasch, J. D. Jost, I. Mirgorodskiy, G. Lihachev, M. L. Gorodetsky, and T. J. Kippenberg, "Mode spectrum and temporal soliton formation in optical microresonators," *Phys. Rev. Lett.*, vol. 113, no. 12, p. 123901, 2014.

[70] A. Kordts, M. H. P. Pfeiffer, H. Guo, V. Brasch, and T. J. Kippenberg, "Higher order mode suppression in high-Q anomalous dispersion SiN microresonators for temporal dissipative Kerr soliton formation," *Opt. Lett.*, vol. 41, no. 3, pp. 452–455, 2016.

[71] T. Herr, V. Brasch, J. J. D., C. Y. Wang, N. M. Kondratiev, M. L. Gorodetsky, and T. J. Kippenberg, "Temporal solitons in optical microresonators," *Nature Photon.*, vol. 8, no. 2, pp. 145–152, 2014.

[72] T. J. Kippenberg, R. Holzwarth, and S. A. Diddams, "Microresonator-based optical frequency combs.," *Science*, vol. 332, no. 6029, pp. 555–559, 2011.

[73] N. Lindenmann, G. Balthasar, D. Hillerkuss, R. Schmogrow, M. Jordan, J. Leuthold, W. Freude, and C. Koos, "Photonic wire bonding: A novel concept for chip-scale interconnects," *Opt. Express*, vol. 20, no. 16, pp. 17667–17677, 2012.

[74] J. S. Levy, A. Gondarenko, M. a. Foster, A. C. Turner-Foster, A. L. Gaeta, and M. Lipson, "CMOS-compatible multiple-wavelength oscillator for on-chip optical interconnects," *Nature Photon.*, vol. 4, no. 1, pp. 37–40, 2010.

[75] A. Liu, L. Liao, Y. Chetrit, J. Basak, H. Nguyen, D. Rubin, and M. Paniccia, "Wavelength Division Multiplexing Based Photonic Integrated Circuits on Silicon-on-Insulator Platform," *IEEE J. Sel. Top. Quant.*, vol. 16, no. 1, pp. 23–32, 2010.

[76] P. Dong, L. Chen, C. Xie, L. L. Buhl, and Y.-K. Chen, "50-Gb/s silicon quadrature phase-shift keying modulator," *Opt. Express*, vol. 20, no. 19, pp. 21181–21186, 2012.

[77] D. Korn, R. Palmer, H. Yu, P. C. Schindler, L. Alloatti, M. Baier, R. Schmogrow, W. Bogaerts, S. K. Selvaraja, G. Lepage, M. Pantouvaki, J. M. D. Wouters, P. Verheyen, J. Van Campenhout, B. Chen, R. Baets, P. Absil, R. Dinu, C. Koos, W. Freude, and J. Leuthold, "Silicon-organic hybrid (SOH) IQ modulator using the linear electro-optic effect for transmitting 16QAM at 112 Gbit/s," *Opt. Express*, vol. 21, no. 11, pp. 13219–13227, 2013.

[78] S. A. Diddams, "The evolving optical frequency comb," *J. Opt. Soc. Am. B*, vol. 27, no. 11, pp. B51–B62, 2010.

[79] T. Hänsch, "Nobel Lecture: Passion for precision," *Rev. Mod. Phys.*, vol. 78, no. 4, pp. 1297–1309, 2006.

[80] J. L. Hall, "Nobel Lecture: Defining and measuring optical frequencies," *Rev. Mod. Phys.*, vol. 78, no. 4, pp. 1279–1295, 2006.

[81] T. Udem, R. Holzwarth, and T. W. Hansch, "Optical frequency metrology," *Nature*, vol. 416, no. 6877, pp. 233–237, 2002.

[82] J. Cundiff, Steven T. and Ye, Ed., *Femtosecond Optical Frequency Comb: Principle, Operation, and Applications*. 2005.

[83] W. Sibbett, A. A. Lagatsky, and C. T. A. Brown, "The development and application of femtosecond laser systems," *Opt. Express*, vol. 20, no. 7,

pp. 6989–7001, 2012.

[84] A. B. Grudinin and S. Gray, "Passive harmonic mode locking in soliton fiber lasers," *J. Opt. Soc. Am. B*, vol. 14, no. 1. p. 144, 1997.

[85] U. Keller, K. J. Weingarten, F. X. Kärtner, D. Kopf, B. Braun, I. D. Jung, R. Fluck, C. Hönninger, N. Matuschek, and J. Aus Der Au, "Semiconductor saturable absorber mirrors (SESAM's) for femtosecond to nanosecond pulse generation in solid-state lasers," *IEEE J. Sel. Top. Quant.*, vol. 2, no. 3, pp. 435–451, 1996.

[86] A. E. H. Oehler, T. Suedmeyer, K. J. Weingarten, and U. Keller, "100 GHz passively mode-locked Er:Yb:glass laser at 1.5 μm with 1.6-ps pulses," *Opt. Express*, vol. 16, no. 26, pp. 21930–21935, 2008.

[87] M. Y. Sander, S. Frolov, J. Shmulovich, E. P. Ippen, and F. X. Kärtner, "10 GHz femtosecond pulse interleaver in planar waveguide technology," *Opt. Express*, vol. 20, no. 4, pp. 4102–4113, 2012.

[88] A. Martinez and S. Yamashita, "Multi-gigahertz repetition rate passively modelocked fiber lasers using carbon nanotubes," *Opt. Express*, vol. 19, no. 7, pp. 6155–6163, 2011.

[89] D. J. Derickson, R. J. Helkey, A. Mar, J. R. Karin, J. G. Wasserbauer, and J. E. Bowers, "Short pulse generation using multisegment mode-locked semiconductorlasers," *IEEE J. Quantum Elect.*, vol. 28, no. 10, 1992.

[90] E. A. Avrutin, J. H. Marsh, and E. L. Portnoi, "Monolithic and multi-GigaHertz mode-locked semiconductor lasers: Constructions, experiments, models and applications," *IEE Proc. - J*, vol. 147, no. 4, pp. 251–278, 2000.

[91] B. R. Koch, A. W. Fang, O. Cohen, and J. E. Bowers, "Mode-locked silicon evanescent lasers," *Opt. Express*, vol. 15, no. 18, pp. 11225–11233, 2007.

[92] V. Torres-Company and A. M. Weiner, "Optical frequency comb technology for ultra-broadband radio-frequency photonics," *Laser Photonics Rev.*, vol. 8, no. 3, pp. 368–393, 2014.

[93] S. Ozharar, F. Quinlan, I. Ozdur, S. Gee, and P. J. Delfyett, "Ultraflat Optical Comb Generation by Phase-Only Modulation of Continuous-Wave Light," *IEEE Photonic. Tech. L.*, vol. 20, no. 1, pp. 36–38, 2008.

[94] M. Fujiwara, M. Teshima, J. Kani, H. Suzuki, N. Takachio, and K. Iwatsuki, "Optical carrier supply module using flattened optical multicarrier generation based on sinusoidal amplitude and phase hybrid modulation," *J. Lightwave Technol.*, vol. 21, no. 11, pp. 2705–2714,

2003.

[95] T. Sakamoto, T. Kawanishi, and M. Izutsu, "Widely wavelength-tunable ultra-flat frequency comb generation using conventional dual-drive Mach-Zehnder modulator," *Electron. Lett.*, vol. 43, no. 19, pp. 1039–1040, 2007.

[96] T. Sakamoto, T. Kawanishi, and M. Izutsu, "Asymptotic formalism for ultraflat optical frequency comb generation using a Mach-Zehnder modulator," *Opt. Lett.*, vol. 32, no. 11, pp. 1515–1517, 2007.

[97] I. L. Gheorma and G. K. Gopalakrishnan, "Flat frequency comb generation with an integrated dual-parallel modulator," *IEEE Photonic. Tech. L.*, vol. 19, no. 13, pp. 1011–1013, 2007.

[98] T. Yamamoto, K. Hitomi, W. Kobayashi, and H. Yasaka, "Optical Frequency Comb Block Generation by Using Semiconductor Mach Zehnder Modulator," *IEEE Photonic. Tech. L.*, vol. 25, no. 1, pp. 40–42, 2013.

[99] R. Slavik, S. G. Farwell, M. J. Wale, and D. J. Richardson, "Compact Optical Comb Generator Using InP Tunable Laser and Push-Pull Modulator," *IEEE Photonic. Tech. L.*, vol. 27, no. 2, pp. 217–220, 2015.

[100] P. M. Anandarajah, R. Zhou, R. Maher, D. M. G. Pascual, F. Smyth, V. Vujicic, and L. Barry, "Flexible Optical Comb Source for Super Channel Systems," in *Optical Fiber Communication Conference*, 2013, p. OTh3I.8.

[101] R. Wu, V. R. Supradeepa, C. M. Long, D. E. Leaird, and A. M. Weiner, "Generation of very flat optical frequency combs from continuous-wave lasers using cascaded intensity and phase modulators driven by tailored radio frequency waveforms," *Opt. Lett.*, vol. 35, no. 19, pp. 3234–3236, 2010.

[102] A. J. Metcalf, V. Torres-Company, D. E. Leaird, and A. M. Weiner, "High-Power Broadly Tunable Electrooptic Frequency Comb Generator," *IEEE J. Quantum Elect.*, vol. 19, no. 6, pp. 231–236, 2013.

[103] F. Tian, X. Zhang, J. Li, and L. Xi, "Generation of 50 Stable Frequency-Locked Optical Carriers for Tb/s Multicarrier Optical Transmission Using a Recirculating Frequency Shifter," *J. Lightwave Technol.*, vol. 29, no. 8, pp. 1085–1091, 2011.

[104] J. Zhang, N. Chi, and J. Yu, "Generation of coherent and frequency-lock multi-carriers using cascaded phase modulators and recirculating frequency shifter for Tb/s optical communication," *Opt. Express*, vol. 19, no. 14, pp. 12891–12902, 2011.

[105] N. Dupuis, C. R. Doerr, L. Zhang, L. Chen, N. J. Sauer, P. Dong, L. L. Buhl, and D. Ahn, "InP-Based Comb Generator for Optical OFDM," *J. Lightwave Technol.*, vol. 30, no. 4, pp. 466–472, 2012.

[106] G. P. Agrawal, *Nonlinear fiber optics*, 4. ed. Amsterdam: Elsevier Academic Press, 2007.

[107] R. W. Boyd, *Nonlinear optics*, 3. ed. Amsterdam: Elsevier, Academic Press, 2008.

[108] Y. R. Shen, *The principles of nonlinear optics*, Wiley clas. Hoboken, NJ: Wiley-Interscience, 2003.

[109] Z. Tong, A. O. J. Wiberg, E. Myslivets, B. P. P. Kuo, N. Alic, and S. Radic, "Spectral linewidth preservation in parametric frequency combs seeded by dual pumps," *Opt. Express*, vol. 20, no. 16, pp. 17610–17619, 2012.

[110] V. Ataie, E. Myslivets, B. P.-P. Kuo, N. Alic, and S. Radic, "Spectrally Equalized Frequency Comb Generation in Multistage Parametric Mixer With Nonlinear Pulse Shaping," *J. Lightwave Technol.*, vol. 32, no. 4, pp. 840–846, 2014.

[111] A. E. H. Oehler, S. C. Zeller, K. J. Weingarten, and U. Keller, "Broad multiwavelength source with 50 GHz channel spacing for wavelength division multiplexing applications in the telecom C band," *Opt. Lett.*, vol. 33, no. 18, pp. 2158–2160, 2008.

[112] D. Hillerkuss, T. Schellinger, M. Jordan, C. Weimann, F. Parmigiani, B. Resan, K. Weingarten, S. Ben-Ezra, B. Nebendahl, C. Koos, W. Freude, and J. Leuthold, "High-Quality Optical Frequency Comb by Spectral Slicing of Spectra Broadened by SPM," *IEEE Photon. J.*, vol. 5, no. 5, p. 7201011, 2013.

[113] A. H. Gnauck, B. P. P. Kuo, E. Myslivets, R. M. Jopson, M. Dinu, J. E. Simsarian, P. J. Winzer, and S. Radic, "Comb-Based 16-QAM Transmitter Spanning the C and L Bands," *IEEE Photonic. Tech. L.*, vol. 26, no. 8, pp. 821–824, 2014.

[114] R. Schmogrow, B. Nebendahl, M. Winter, A. Josten, D. Hillerkuss, S. Koenig, J. Meyer, M. Dreschmann, M. Huebner, C. Koos, J. Becker, W. Freude, and J. Leuthold, "Error Vector Magnitude as a Performance Measure for Advanced Modulation Formats," *IEEE Photonic. Tech. L.*, vol. 24, no. 1, pp. 61–63, 2012.

[115] R. Schmogrow, B. Nebendahl, M. Winter, A. Josten, D. Hillerkuss, S. Koenig, J. Meyer, M. Dreschmann, M. Huebner, C. Koos, J. Becker, W. Freude, and J. Leuthold, "Corrections to 'Error Vector Magnitude as a

Performance Measure for Advanced Modulation Formats' [Jan 1, 2012 61-63]," *IEEE Photonic. Tech. L.*, vol. 24, no. 23, p. 2198, 2012.

[116] M. Nakazawa, T. Hirooka, M. Yoshida, and K. Kasai, "Extremely Higher-Order Modulation Format," in *OPTICAL FIBER TELECOMMUNICATIONS VIB: SYSTEMS AND NETWORKS*, 6th ed., I. Kaminow, T. Li, and A. E. Willner, Eds. Academic Press, 2013, pp. 297–336.

[117] G. P. Agrawal, *Lightwave technology*, vol. [1]: Compo. Hoboken, NJ: Wiley-Interscience, 2004.

[118] E. Ip and J. M. Kahn, "Feedforward Carrier Recovery for Coherent Optical Communications," *J. Lightwave Technol.*, vol. 25, no. 9, pp. 2675–2692, 2007.

[119] T. Pfau, S. Hoffmann, and R. Noe, "Hardware-efficient coherent digital receiver concept with feedforward carrier recovery for M-QAM constellations," *J. Lightwave Technol.*, vol. 27, no. 8, pp. 989–999, 2009.

[120] S. Zhang, L. Xu, P. Y. Kam, C. Yu, J. Chen, and T. Wang, "A Performance Investigation of Adaptive Phase Estimations in Coherent Optical Communications," *IEEE Photonic. Tech. L.*, vol. 23, no. 8, pp. 462–464, 2011.

[121] R. Hui and M. S. O'Sullivan, *Fiber optic measurement techniques*. Amsterdam: Elsevier Academic Press, 2009.

[122] T. Okoshi, K. Kikuchi, and A. Nakayama, "Novel method for high resolution measurement of laser output spectrum," *Electron. Lett.*, vol. 16, no. 16, pp. 630–631, 1980.

[123] S. Gringeri, E. B. Basch, and T. J. Xia, "Technical considerations for supporting data rates beyond 100 Gb/s," *IEEE Commun. Mag.*, vol. 50, no. 2, pp. s21–s30, 2012.

[124] W. Shieh and I. Djordjevic, *OFDM for optical communications*. Amsterdam: Elsevier Academic Press, 2010.

[125] H.-C. Chien, J. Yu, Z. Jia, Z. Dong, and X. Xiao, "Performance Assessment of Noise-Suppressed Nyquist-WDM for Terabit Superchannel Transmission," *J. Lightwave Technol.*, vol. 30, no. 24, pp. 3965–3971, 2012.

[126] G. Bosco, V. Curri, A. Carena, P. Poggiolini, and F. Forghieri, "On the Performance of Nyquist-WDM Terabit Superchannels Based on PM-BPSK, PM-QPSK, PM-8QAM or PM-16QAM Subcarriers," *J. Lightwave Technol.*, vol. 29, no. 1, pp. 53–61, 2011.

[127] Y.-K. Huang, E. Mateo, M. Sato, D. Qian, F. Yaman, T. Inoue, Y. Inada, S. Zhang, Y. Aono, T. Tajima, T. Ogata, and Y. Aoki, "Real-time Transoceanic Transmission of 1-Tb/s Nyquist Superchannel at 2.86-b/s/Hz Spectral Efficiency," in *Asia Communications and Photonics Conference*, 2012, p. PAF4C.2.

[128] X. Liu, D. M. Gill, S. Chandrasekhar, L. L. Buhl, M. Earnshaw, M. a. Cappuzzo, L. T. Gomez, Y. Chen, F. P. Klemens, E. C. Burrows, Y. K. Chen, and R. W. Tkach, "Multi-carrier coherent receiver based on a shared optical hybrid and a cyclic AWG array for terabit/s optical transmission," *IEEE Photon. J.*, vol. 2, no. 3, pp. 330–337, 2010.

[129] X. Liu, S. Chandrasekhar, X. Chen, P. J. Winzer, Y. Pan, T. F. Taunay, B. Zhu, M. Fishteyn, M. F. Yan, J. M. Fini, E. M. Monberg, and F. V Dimarcello, "1.12-Tb/s 32-QAM-OFDM superchannel with 8.6-b/s/Hz intrachannel spectral efficiency and space-division multiplexed transmission with 60-b/s/Hz aggregate spectral efficiency," *Opt. Express*, vol. 19, no. 26, pp. B958--B964, 2011.

[130] E. Temprana, V. Ataie, B. P.-P. Kuo, E. Myslivets, N. Alic, and S. Radic, "Low-noise parametric frequency comb for continuous C-plus-L-band 16-QAM channels generation," *Opt. Express*, vol. 22, no. 6, pp. 6822–6828, 2014.

[131] A. K. Mishra, R. Schmogrow, I. Tomkos, D. Hillerkuss, C. Koos, W. Freude, and J. Leuthold, "Flexible RF-Based Comb Generator," *IEEE Photonic. Tech. L.*, vol. 25, no. 7, pp. 701–704, 2013.

[132] P. M. Anandarajah, R. Zhou, R. Maher, D. Lavery, M. Paskov, B. Thomsen, S. Savory, and L. P. Barry, "Gain-switched multicarrier transmitter in a long-reach UDWDM PON with a digital coherent receiver," *Opt. Lett.*, vol. 38, no. 22, p. 4797, 2013.

[133] R. Zhou, P. M. Anandarajah, R. Maher, M. Paskov, D. Lavery, B. C. Thomsen, S. J. Savory, and L. P. Barry, "80-km Coherent DWDM-PON on 20-GHz Grid With Injected Gain Switched Comb Source," *IEEE Photonic. Tech. L.*, vol. 26, no. 4, pp. 364–367, 2014.

[134] R. Zhou, T. N. Huynh, V. Vujicic, P. M. Anandarajah, and L. P. Barry, "Phase noise analysis of injected gain switched comb source for coherent communications," *Opt. Express*, vol. 22, no. 7, pp. 8120–8125, 2014.

[135] R. Zhou, P. M. Anandarajah, D. G. Pascual, J. O'Carroll, R. Phelan, B. Kelly, and L. P. Barry, "Monolithically Integrated 2-Section Lasers for Injection Locked Gain Switched Comb Generation," in *Optical Fiber*

Communication Conference, 2014, p. Th3A.3.

[136] R. Schmogrow, D. Hillerkuss, M. Dreschmann, M. Huebner, M. Winter, J. Meyer, B. Nebendahl, C. Koos, J. Becker, W. Freude, and J. Leuthold, "Real-Time Software-Defined Multiformat Transmitter Generating 64QAM at 28 GBd," *IEEE Photonic. Tech. L.*, vol. 22, no. 21, pp. 1601–1603, 2010.

[137] P. Winzer, A. Gnauck, A. Konczykowska, F. Jorge, and J.-Y. Dupuy, "Penalties from in-band crosstalk for advanced optical modulation formats," in *37th European Conference and Exposition on Optical Communications*, 2011, p. Tu.5.B.7.

[138] F. Chang, K. Onohara, and T. Mizuochi, "Forward error correction for 100 G transport networks," *IEEE Commun. Mag.*, vol. 48, no. 3, pp. S48–S55, 2010.

[139] B. Li, K. J. Larsen, D. Zibar, and I. T. Monroy, "Over 10 dB Net Coding Gain Based on 20% Overhead Hard Decision Forward Error Correction in 100G Optical Communication Systems," in *37th European Conference and Exposition on Optical Communications*, 2011, p. Tu.6.A.3.

[140] J. Renaudier, R. Brenot, B. Dagens, F. Lelarge, B. Rousseau, F. Poingt, O. Legouezigou, F. Pommereau, A. Accard, P. Gallion, and G.-H. Duan, "45 GHz self-pulsation with narrow linewidth in quantum dot Fabry-Perot semiconductor lasers at 1.5 µm," *Electron. Lett.*, vol. 41, no. 18, pp. 1007–1008, 2005.

[141] C. Gosset, K. Merghem, A. Martinez, G. Moreau, G. Patriarche, G. Aubin, J. Landreau, F. Lelarge, and A. Ramdane, "Sub-picosecond pulse generation at 134 GHz using a quantum dash-based Fabry-Perot laser emitting at 1.56 µm," in *Optical Fiber Communication Conference*, 2006, p. OThG1.

[142] K. Merghem, C. Calo, R. Rosales, X. Lafosse, G. Aubin, A. Martinez, F. Lelarge, and A. Ramdane, "Stability of optical frequency comb generated with InAs/InP quantum-dash-based passive mode-locked lasers," *IEEE J. Quantum Elect.*, vol. 50, no. 4, pp. 275–280, 2014.

[143] T. Habruseva, S. O'Donoghue, N. Rebrova, F. Kéfélian, S. P. Hegarty, and G. Huyet, "Optical linewidth of a passively mode-locked semiconductor laser," *Opt. Lett.*, vol. 34, no. 21, pp. 3307–3309, 2009.

[144] R. Rosales, K. Merghem, A. Martinez, F. Lelarge, A. Accard, and A. Ramdane, "Timing jitter from the optical spectrum in semiconductor passively mode locked lasers," *Opt. Express*, vol. 20, no. 8, pp. 9151–

9160, 2012.

[145] Y. Takushima, H. Sotobayashi, M. E. Grein, E. P. Ippen, and H. A. Haus, "Linewidth of mode combs of passively and actively mode-locked semiconductor laser diodes," *Proc. SPIE*, vol. 5595. pp. 213–227, 2004.

[146] R. Watts, S. G. Murdoch, and L. Barry, "Spectral linewidth reduction of single-mode and mode-locked lasers using a feed-forward heterodyne detection scheme," in *CLEO: 2014*, 2014, p. STh3O.8.

[147] Y. Ben M'Sallem, Q. T. Le, L. Bramerie, Q. Nguyen, E. Borgne, P. Besnard, A. Shen, F. Lelarge, S. LaRochelle, L. A. Rusch, and J. Simon, "Quantum-dash mode-locked laser as a source for 56-Gb/s DQPSK modulation in WDM multicast applications," *IEEE Photonic. Tech. L.*, vol. 23, no. 7, pp. 453–455, 2011.

[148] G. P. Agrawal and N. K. Dutta, *Semicondutctor Lasers*, 2nd ed. Van Nostrand Reinhold, New York, 1993.

[149] R. Watts, R. Rosales, F. Lelarge, A. Ramdane, and L. Barry, "Mode coherence measurements across a 1.5 THz spectral bandwidth of a passively mode-locked quantum dash laser," *Opt. Lett.*, vol. 37, no. 9, pp. 1499–1501, 2012.

[150] A. Akrout, A. Shen, R. Brenot, F. Van Dijk, O. Legouezigou, F. Pommereau, F. Lelarge, A. Ramdane, and G.-H. Duan, "Separate error-free transmission of eight channels at 10 Gb/s using comb generation in a quantum-dash-based mode-locked laser," *IEEE Photonic. Tech. L.*, vol. 21, no. 23, pp. 1746–1748, 2009.

[151] E. Sooudi, G. Huyet, J. G. McInerney, F. Lelarge, K. Merghem, R. Rosales, A. Martinez, A. Ramdane, and S. P. Hegarty, "Injection-locking properties of InAs/InP-based mode-locked quantum-dash lasers at 21 GHz," *IEEE Photonic. Tech. L.*, vol. 23, no. 20, pp. 1544–1546, 2011.

[152] T. Pfau, R. Peveling, J. Hauden, N. Grossard, H. Porte, Y. Achiam, S. Hoffmann, S. K. Ibrahim, O. Adamczyk, S. Bhandare, D. Sandel, M. Porrmann, and R. Noe, "Coherent digital polarization diversity receiver for real-time polarization-multiplexed QPSK transmission at 2.8 Gb/s," *IEEE Photonic. Tech. L.*, vol. 19, no. 24, pp. 1988–1990, 2007.

[153] D. Lavery, R. Maher, M. Paskov, B. C. Thomsen, P. Bayvel, and S. J. Savory, "Digital coherence enhancement enabling 6-GBd DP-64QAM using a 1.4-MHz linewidth laser," *IEEE Photonic. Tech. L.*, vol. 25, no. 22, pp. 2213–2216, 2013.

[154] D. Derickson, "Fiber optic test and measurement," in *Fiber optic test and*

measurement/edited by Dennis Derickson. Upper Saddle River, NJ: Prentice Hall, 1998, vol. 1.

[155] B. Puttnam, R. Luís, J. Delgado Mendinueta, J. Sakaguchi, W. Klaus, Y. Kamio, M. Nakamura, N. Wada, Y. Awaji, A. Kanno, T. Kawanishi, and T. Miyazaki, "Self-homodyne detection in optical communication systems," *Photonics*, vol. 1, no. 2, pp. 110–130, 2014.

[156] A. A. Savchenkov, A. B. Matsko, V. S. Ilchenko, I. Solomatine, D. Seidel, and L. Maleki, "Tunable optical frequency comb with a crystalline whispering gallery mode resonator," *Phys. Rev. Lett.*, vol. 101, no. 9, p. 93902, 2008.

[157] W. Liang, A. A. Savchenkov, A. B. Matsko, V. S. Ilchenko, D. Seidel, and L. Maleki, "Generation of near-infrared frequency combs from a MgF2 whispering gallery mode resonator," *Opt. Lett.*, vol. 36, no. 12, pp. 2290–2292, 2011.

[158] D. Braje, L. Hollberg, and S. Diddams, "Brillouin-enhanced hyperparametric generation of an optical frequency comb in a monolithic highly nonlinear fiber cavity pumped by a cw laser," *Phys. Rev. Lett.*, vol. 102, no. 19, p. 193902, 2009.

[159] L. Razzari, D. Duchesne, M. Ferrera, R. Morandotti, S. Chu, B. E. Little, and D. J. Moss, "CMOS-compatible integrated optical hyper-parametric oscillator," *Nature Photon.*, vol. 4, no. 1, pp. 41–45, 2010.

[160] J. Li, H. Lee, T. Chen, and K. J. Vahala, "Low-pump-power, low-phase-noise, and microwave to millimeter-wave repetition rate operation in microcombs," *Phys. Rev. Lett.*, vol. 109, no. 23, p. 233901, 2012.

[161] B. J. M. Hausmann, I. Bulu, V. Venkataraman, P. Deotare, and M. Lončar, "Diamond nonlinear photonics," *Nature Photon.*, vol. 8, pp. 1–6, 2014.

[162] H. Jung, C. Xiong, K. Y. Fong, X. Zhang, and H. X. Tang, "Optical frequency comb generation from aluminum nitride microring resonator," *Opt. Lett.*, vol. 38, no. 15, pp. 2810–2813, 2013.

[163] A. Schliesser, N. Picqué, and T. W. Hänsch, "Mid-infrared frequency combs," *Nature Photon.*, vol. 6. pp. 440–449, 2012.

[164] X. Zhang, T. Liu, J. Jiang, M. Feng, and K. Liu, "Mid-infrared frequency comb generation in coupled silicon microring resonators," *Opt. Commun.*, vol. 332, pp. 125–131, 2014.

[165] L. Zhang, Y. Yue, R. G. Beausoleil, and A. E. Willner, "Analysis and engineering of chromatic dispersion in silicon waveguide bends and ring resonators," *Opt. Express*, vol. 19, no. 9, pp. 8102–8107, 2011.

[166] D. Miller, "Device requirements for optical interconnects to silicon chips," *Proc. IEEE*, vol. 97, no. 7, pp. 1166–1185, 2009.

[167] D. Feng, W. Qian, H. Liang, B. J. Luff, and M. Asghari, "High-speed receiver technology on the SOI platform," *IEEE J. Sel. Top. Quant.*, vol. 19, no. 2, p. 3800108, 2013.

[168] J. S. Levy, K. Saha, Y. Okawachi, M. A. Foster, A. L. Gaeta, and M. Lipson, "High-Performance Silicon-Nitride-Based Multiple-Wavelength Source," *IEEE Photonic. Tech. L.*, vol. 24, no. 16, pp. 1375–1377, 2012.

[169] T. J. Kippenberg, S. M. Spillane, and K. J. Vahala, "Kerr-nonlinearity optical parametric oscillation in an ultrahigh-Q toroid microcavity," *Phys. Rev. Lett.*, vol. 93, no. 8, p. 83904, 2004.

[170] A. Coillet, I. Balakireva, R. Henriet, K. Saleh, L. Larger, J. M. Dudley, C. R. Menyuk, and Y. K. Chembo, "Azimuthal Turing Patterns, Bright and Dark Cavity Solitons in Kerr Combs Generated With Whispering-Gallery-Mode Resonators," *IEEE Photon. J.*, vol. 5, no. 4, p. 6100409, 2013.

[171] A. Coillet and Y. Chembo, "On the robustness of phase locking in Kerr optical frequency combs," *Opt. Lett.*, vol. 39, no. 6, p. 1529, 2014.

[172] A. M. Turing, "The Chemical Basis of Morphogenesis," *Philos. T. Roy. Soc. B*, vol. 237, no. 641, pp. 37–72, 1952.

[173] C. Godey, I. V Balakireva, A. Coillet, and Y. K. Chembo, "Stability analysis of the spatiotemporal Lugiato-Lefever model for Kerr optical frequency combs in the anomalous and normal dispersion regimes," *Phys. Rev. A*, vol. 89, no. 6, p. 63814, 2014.

[174] P. H. Wang, F. Ferdous, H. X. Miao, J. Wang, D. E. Leaird, K. Srinivasan, L. Chen, V. Aksyuk, and A. M. Weiner, "Observation of correlation between route to formation, coherence, noise, and communication performance of Kerr combs," *Opt. Express*, vol. 20, no. 28, pp. 29284–29295, 2012.

[175] Y. Okawachi, K. Saha, J. S. Levy, Y. H. Wen, M. Lipson, and A. L. Gaeta, "Octave-spanning frequency comb generation in a silicon nitride chip," *Opt. Lett.*, vol. 36, no. 17, pp. 3398–3400, 2011.

[176] J. Li, K. Worms, R. Maestle, D. Hillerkuss, W. Freude, and J. Leuthold, "Free-space optical delay interferometer with tunable delay and phase," *Opt. Express*, vol. 19, no. 12, pp. 11654–11666, 2011.

[177] P.-H. Wang, Y. Xuan, L. Fan, L. T. Varghese, J. Wang, Y. Liu, X. Xue, D. E. Leaird, M. Qi, and A. M. Weiner, "Drop-port study of microresonator frequency combs: Power transfer, spectra and time-domain

characterization," *Opt. Express*, vol. 21, no. 19, pp. 22441–22452, 2013.

[178] M. H. P. Pfeiffer, A. Kordts, V. Brasch, M. Zervas, M. Geiselmann, J. D. Jost, and T. J. Kippenberg, "Photonic Damascene process for integrated high-Q microresonator based nonlinear photonics," *Optica*, vol. 3, no. 1, pp. 20–25, 2016.

[179] ITU-T Recommendation G.975.1, "Forward error correction for high bit-rate DWDM submarine systems." 2004.

[180] C. Bao, L. Zhang, A. Matsko, Y. Yan, Z. Zhao, G. Xie, A. M. Agarwal, L. C. Kimerling, J. Michel, L. Maleki, and A. E. Willner, "Nonlinear conversion efficiency in Kerr frequency comb generation," *Opt. Lett.*, vol. 39, no. 21, pp. 6126–6129, 2014.

[181] V. E. Lobanov, G. Lihachev, T. J. Kippenberg, and M. L. Gorodetsky, "Frequency combs and platicons in optical microresonators with normal GVD," *Opt. Express*, vol. 23, no. 6, pp. 7713–7721, 2015.

[182] V. Brasch, M. Geiselmann, T. Herr, G. Lihachev, M. H. P. Pfeiffer, M. L. Gorodetsky, and T. J. Kippenberg, "Photonic chip-based optical frequency comb using soliton Cherenkov radiation," *Science*, vol. 351, no. 6271, pp. 357–360, 2016.

[183] Y. K. Chembo and N. Yu, "Modal expansion approach to optical-frequency-comb generation with monolithic whispering-gallery-mode resonators," *Phys. Rev. A*, vol. 82, no. 3, p. 33801, 2010.

[184] J. M. Dudley, G. Genty, and S. Coen, "Supercontinuum generation in photonic crystal fiber," *Rev. Mod. Phys.*, vol. 78, no. 4, pp. 1135–1184, 2006.

[185] T. Carmon, L. Yang, and K. Vahala, "Dynamical thermal behavior and thermal self-stability of microcavities," *Opt. Express*, vol. 12, no. 20, pp. 4742–4750, 2004.

[186] J. M. Kahn and K.-P. Ho, "Spectral efficiency limits and modulation/detection techniques for DWDM systems," *IEEE J. Sel. Top. Quantum Electron.*, vol. 10, no. 2, pp. 259–272, 2004.

[187] P. J. Winzer and R. Essiambre, "Advanced Optical Modulation Formats," *Proc. IEEE*, vol. 94, no. 5, pp. 952–985, 2006.

[188] R. Schmogrow, S. Ben-Ezra, P. C. Schindler, B. Nebendahl, C. Koos, W. Freude, and J. Leuthold, "Pulse-Shaping With Digital, Electrical, and Optical Filters—A Comparison," *J. Lightwave Technol.*, vol. 31, no. 15, pp. 2570–2577, 2013.

[189] C. Herard and A. Lacourt, “New multiplexing technique using polarization of light,” *Appl. Opt.*, vol. 30, no. 2, pp. 222–231, 1991.

[190] B. Szafraniec, B. Nebendahl, and T. Marshall, “Polarization demultiplexing in Stokes space,” *Opt. Express*, vol. 18, no. 17, pp. 17928–17939, 2010.

[191] R. Schmogrow, P. C. Schindler, C. Koos, W. Freude, and J. Leuthold, “Blind Polarization Demultiplexing With Low Computational Complexity,” *IEEE Photonic. Tech. L.*, vol. 25, no. 13, pp. 1230–1233, 2013.

[192] ITU-T Recommendation G.652, “Characteristics of a single-mode optical fibre and cable.” 2009.

[193] F. Derr, “Coherent optical QPSK intradyne system: concept and digital receiver realization,” *J. Lightwave Technol.*, vol. 10, no. 9, pp. 1290–1296, 1992.

[194] S. J. Savory, “Digital Coherent Optical Receivers: Algorithms and Subsystems,” *IEEE J. Sel. Top. Quant.*, vol. 16, no. 5, pp. 1164–1179, 2010.

[195] E. M. Ip and J. M. Kahn, “Fiber Impairment Compensation Using Coherent Detection and Digital Signal Processing,” *J. Lightwave Technol.*, vol. 28, no. 4, pp. 502–519, 2010.

[196] K. Onohara, T. Sugihara, Y. Miyata, K. Sugihara, K. Kubo, H. Yoshida, K. Koguchi, and T. Mizuochi, “Soft-decision forward error correction for 100Gb/s digital coherent systems,” *Opt. Fiber Technol.*, vol. 17, no. 5, pp. 452–455, 2011.

[197] C. Jeruchim, “Techniques for Estimating the Bit Error Rate in the Simulation of Digital Communication Systems,” *IEEE J. Sel. Area. Comm.*, vol. 2, no. 1, pp. 153–170, 1984.

[198] W. Freude, R. Schmogrow, B. Nebendahl, M. Winter, A. Josten, D. Hillerkuss, S. Koenig, J. Meyer, M. Dreschmann, M. Huebner, C. Koos, J. Becker, and J. Leuthold, “Quality metrics for optical signals: Eye diagram, Q-factor, OSNR, EVM and BER,” in *International Conference on Transparent Optical Networks (ICTON)*, 2012, p. Mo.B1.5.

[199] M. J. E. Golay, “Note on coherence vs narrow-bandedness in regenerative oscillators, masers, lasers, etc.,” *Proc. IRE*, vol. 49, pp. 958–959, 1961.

[200] R. Adler, “A Study of Locking Phenomena in Oscillators,” *Proc. IRE*, vol. 34, no. 6, pp. 351–357, 1946.

[201] C. H. Henry, "Theory of the linewidth of semiconductor lasers," *IEEE J. Quantum Elect.*, vol. 18, no. 2, pp. 259–264, 1982.

[202] W. Freude, "Optical Transmitters and Receivers." Lecture Notes, Institute of Photoncs and Quantum Electronics, Karlsruhe Institute of Technology (KIT), 2014.

[203] J. Poette, P. Besnard, L. Bramerie, and J.-C. Simon, "Highly-sensitive measurement technique of relative intensity noise and laser characterization," *Proc. SPIE*, vol. 6603. p. 66031R–66031R–10, 2007.

Appendix E Glossary

E.1 List of symbols

E.1.1 Greek symbols

α	Attenuation coefficient
α_H	Henry factor
β	Propagation constant
$\beta_c^{(i)}$	Taylor-coefficients for the propagation constant
$\chi^{(n)}$	Susceptibility of n-th order
Δ, δ	Spacing of primary and secondary comb lines
$\Delta\phi, \Delta\phi_{\mathrm{DC}}$	Phase shifts induced by driving and bias voltages for a MZM
$\delta\lambda$	Cold cavity detuning
$\Delta\tau_r$	Timing jitter of intracavity pulses
ε_0	Electric permittivity of vacuum
η	Photodiode quantum efficiency
λ	Wavelength
ω	Angular frequency
ϕ_m	Static phase of comb line with index m
φ	Phase
φ_{inc}	Phase increments
σ_1, σ_0	Standard deviation of voltage representing logical '1' and '0'
σ_θ^2	Phase noise variance
ρ	Correlation coefficient
τ_r	Cavity round trip time
τ_{dly}	Time delay

E.1.2 Latin symbols

A	Amplitude
B	Bandwidth
B_O	Optical bandwidth
B_{RBW}	Resolution bandwidth
B_{ref}	Reference bandwidth for noise power evaluation
c_0	Speed of light in vacuum
C_λ	Dispersion coefficient
d	Diameter of disc resonator
$D_c^{(k)}$	Taylor-coefficients for resonance angular frequencies

E	Electric field
e	Elementary charge
erfc	Complementary error function
f	Frequency
f_0, f_m	Resonance frequencies
f_r	Free spectral range
$h(\hbar)$	(Modified) Planck's constant
i	Current
I_{th}	Laser diode threshold current
k	Factor accounting for the EVM normalizations
k_B	Boltzmann constant
L	Length
M	Number of symbols or constellation size
N_p	Number of photons
n_e	Effective index of a waveguide
n_{eg}	Effective group index of a waveguide
P	Dielectric polarization induced by an electric field
p	1 or 2 for single polarization or polarization multiplexed signal
$p(0t), p(1t)$	Occurrence probability for transmitted '0' and '1'
P_e	External optical power incident on a photodiode
P_m	Power of comb line with index m
P_N	Noise power
$P_{\text{tot}}, P_{\text{comb}}$	Total chip output power and power in the generated comb lines
Q	Quality factor
$Q_{\text{ext}}, Q_{\text{int}}$	Extrinsic and intrinsic resonator quality factor
q	Oversampling factor
R_L	Load resistance
R_s	Symbol rate
R_{sp}	Rate of spontaneous emission into the lasing mode
S	Receiver sensitivity
S_N	Noise power spectral density
T	Symbol duration
T_r	Temperature of the electrical receiver circuit
t	Time
u	Voltage
v_{eg}	Effective group velocity
v_p	Phase velocity

U_π	Voltage for a π phase difference between the two arms of a MZM
w	Probability density

E.2 Acronyms

16QAM	16-state quadrature amplitude modulation
AC	Alternating current
ADC	Analog-to-digital converter
AR	Anti-reflection
ASE	Amplified spontaneous emission
AWG	Arbitrary waveform generator
B2B	Back-to-back
BER	Bit-error ratio
BERT	Bit error ratio tester
BPF	Band-pass filter
CaF	Calcium fluoride
CAGR	Compound annual growth rate
CMOS	Complementary metal-oxide semiconductor
CW	Continuous wave
DAC	Digital-to-analog converter
DC	Data center
DC	Direct current
DCA	Digital communication analyzer
DCI	Data center interconnect
DEMUX	De-multiplexer
DFB	Distributed feedback
DI	Delay interferometer
DQPSK	Differential quadrature phase shift keying
DSP	Digital signal processing
DWDM	Dense wavelength division multiplexing
ECL	External cavity laser
EDFA	Erbium-doped fiber amplifiers
Eq	Equalizer
ER	Extinction ratio
ESA	Electrical spectrum analyzer
EVM	Error-vector magnitude

EVM_m	Error-vector magnitude normalized to the maximum field in the constellation
FBG	Fiber-Bragg grating
FEC	Forward error correction
FFH	Feed-forward heterodyne
FM	Frequency modulation
FSR	Free-spectral range
FWHM	Full width half maximum
FWM	Four-wave mixing
GSCS	Gain-switched comb source
HNLF	Highly nonlinear fiber
HR	High-reflectivity
HWP	Half-wave plate
I	Inphase
InAs	Indium arsenide
InGaAsP	Indium gallium arsenide phosphide
InP	Indium phosphide
IQ-Mod	In-phase quadrature modulator
ITLA	Integrable tunable laser assembly
ITU	International Telecommunication Union
LED	Light emitting diode
LF	Lensed fiber
LO	Local oscillator
LPCVD	Low-pressure chemical vapor deposition
MgF	Magnesium fluoride
MLL	Mode-locked laser
MSA	Multi-source agreement
MUX	Multiplexer
MZM	Mach-zehnder modulator
NRZ	Non-return-to-zero
OCN_0R	Optical carrier-to-noise-density power ratio
OCNR	Optical carrier-to-noise power ratio
OFDM	Orthogonal frequency-division multiplexing
OMA	Optical modulation analyzer
OOK	On-off keying
OSA	Optical spectrum analyzer
OSNR	Optical signal-to-noise ratio

PBS	Polarization beam splitter
PC	Polarization controller
PD	Photodiode
pdf	Probability density function
PDM	Polarization-division multiplexing
PIC	Photonic integrated circuit
PID	Proportional integral derivative
PM	Power meter
PolMUX	Polarization multiplexing
PON	Passive optical network
PRBS	Pseudo-random bit sequence
PSD	Power spectral density
Q	Quality
Q	Quadrature
QD-MLLD	Quantum-dash mode-locked laser diode
QPSK	Quadrature phase shift keying
RBW	Resolution bandwidth
RF	Radio frequency
RFS	Recirculating frequency shifter
RIN	Relative intensity noise
Rx	Receiver
RZ	Return-to-zero
SE	Spectral Efficiency
SEM	Scanning electron microscope
SESAM	Semiconductor saturable absorption mirror
SiN	Silicon nitride
SiO_2	Silicon dioxide
SNR	Signal-to-noise power ratio
SOH	Silicon-organic hybrid
SPM	Self-phase modulation
SR	Suppression ratio
SSMF	Standard single-mode fibers
TBPF	Tunable band-pass filter
TEC	Thermo-electric cooler
TLS	Tunable laser source
Tx	Transmitter
UVTP	Ultraviolet thermal processing

VCSEL	Vertical cavity surface emitting lasers
VOA	Variable optical attenuator
WDM	Wavelength-division multiplexing
WGR	Whispering gallery mode
WS	Waveshaper
WSS	Wavelength selective switch
XPM	Cross-phase modulation

Acknowledgements (German)

Die vorliegende Dissertation entstand während meiner Tätigkeit am Institut für Photonik und Quantenelektronik (IPQ) am Karlsruher Institut für Technologie (KIT). Sie war eingebunden in das Center for Functional Nanostructures (CFN) der deutschen Forschungsgesellschaft (DFG) (project A 4.8), die von der europäischen Union (EU) geförderten Projekte BIG PIPES (FP7 ICT-STREP, Nummer 619591) und EnTeraPIC (ERC Starting Grant, Nummer 280145), sowie die Karlsruhe School of Optics and Photonics (KSOP). An dieser Stelle möchte ich mich herzlich bei allen Personen bedanken, die mich in den letzten Jahren unterstützt und damit wesentlich zum Gelingen dieser Arbeit beigetragen haben.

Allen voran danke ich meinem Doktorvater Prof. Christian Koos für die Betreuung, die vielen Ideen, die guten Ratschläge, sowie die Bereitstellung der Infrastruktur und der finanziellen Mittel. Besonderer Dank gebührt auch Prof. Wolfgang Freude für die guten Diskussionen, die unermüdlichen Erklärungen und die Durchsicht sämtlicher Manuskripte. Prof. Jeremy Witzens von der Rheinisch-Westfälischen Technischen Hochschule Aachen danke ich herzlich für die Übernahme des Korreferenten und die Kommentare und Anregungen zur Verbesserung der Dissertation. Danke auch an Prof. Jürg Leuthold, der in den ersten Jahren meiner Tätigkeit Institutsleiter des IPQ war und in dieser Zeit meine Arbeit unterstützte.

Die in dieser Arbeit erzielten Ergebnisse wären nicht möglich gewesen ohne die sehr guten Kooperationen mit externen Partnern, die mir die wesentlichen Elemente für die verschiedenen Frequenzkammquellen zur Verfügung gestellt haben und jederzeit für Diskussionen über deren Aufbau und Eigenschaften bereit waren. Darüber hinaus will ich mich auch für die Unterstützung der Publikationen von den jeweiligen Messergebnissen bedanken. Ich danke Prof. Tobias Kippenberg von der Ecole Polytechnique Fédérale de Lausanne und seiner Gruppe für die SiN Mikroresonatoren, sowie Dr. Yanne Chembo vom Femto-ST in Besançon, Frankreich für die gemeinsamen Experimente mit ihren Magnesiumfluorid Mikroresonatoren.

Den BIG PIPES Projektpartnern danke ich für die produktive und fruchtbare Zusammenarbeit. Insbesondere möchte ich hier Prof. Liam Barry und seiner Gruppe an der Dublin City University danken für die gemeinsamen Messkampagnen mit dem „Gain-Switched Comb Source“ als Kammquelle und der Phasenrauschkompensation der Kammlinien eines modengekoppelten

Quantendraht-Lasers. Ein ebenso großer Dank gebührt Prof. Abderrahim Ramdane und seiner Gruppe am Centre National de la Recherche Scientific sowie Dr. Francois Lelarge und seinen Kollegen bei III-V Labs für die Quantendraht-Laserdioden. Ich danke auch dem Projektkoordinator Prof. Jeremy Witzens und seiner Gruppe, sowie den weiteren Projektpartnern von Pilot Photonics Inc. und Mellanox für die gute und unkomplizierte Zusammenarbeit beim Schreiben des Projektantrags, bei den Projekttreffen und generell der projektbezogenen Arbeit.

Ein großes Dankeschön geht an meine Kollegen am IPQ. Insbesondere an das Team im Sekretariat, die immer hilfreich zur Seite standen bei allerlei organisatorischen und administrativen Dingen, sowie die IT Beauftragten und die Laborunterstützung durch das technische Personal, insbesondere Oswald Speck, sowie die Werkstatt, die immer wieder schnelle und gute Lösungen für meine Laboraufbauten anfertigten und auch Sonderwünsche berücksichtigten. Natürlich möchte ich hier auch allen Doktoranden und Postdocs danken, mit denen ich in den letzten Jahren zusammenarbeiten durfte. Neben der Arbeit und den fachlichen Diskussionen kam auch der Spaß nicht zu kurz, ob bei einer Kaffeerunde, am Kickertisch oder beim Grillen. Ich erinnere mich auch gerne an viele gemeinsame Ausflüge zum Beispiel nach Stuttgart ins Stadion des VfB, sowie Roadtrips nach Konferenzen und Sommerschulen.

Für die gute Zusammenarbeit innerhalb des KIT möchte ich mich bedanken bei Prof. Wolfgang Reichel und Prof. Tobias Jahnke aus der Fakultät Mathematik für die lehrreichen Diskussionen zur mathematischen Modellierung und Simulation der Kerr Frequenzkämme. Ferner danke ich den Kollegen vom Institut für Mikrostrukturtechnik (IMT), dem CFN und der KSOP.

Ich bedanke mich bei allen Studenten, die zu dieser Arbeit beigetragen haben: Marcel Dornbusch, Florian Bach, Jorge Mendez, Daniel Wegner, Niels Rademacher, Yimin Yu, Igor Shkarban und Ralf Rosenberger.

Zu guter Letzt danke ich meiner Freundin Barbara, sowie meiner Familie für die jahrelange Unterstützung, die Aufmunterungen und den Zuspruch.

List of Publications

Journal Publications

[J1] **J. Pfeifle**, A. Coillet, R. Henriet, K. Saleh, P. Schindler, C. Weimann, W. Freude, I. V. Balakireva, L. Larger, C. Koos, and Y. K. Chembo, "Optimally coherent Kerr combs generated with crystalline whispering gallery mode resonators for ultrahigh capacity fiber communications," Phys. Rev. Lett. 114, 093902, 2015.

[J2] **J. Pfeifle**, V. Vujicic, R. T. Watts, P. C. Schindler, C. Weimann, R. Zhou, W. Freude, L. P. Barry, and C. Koos, "Flexible terabit/s Nyquist-WDM super-channels using a gain-switched comb source," *Opt. Express*, vol. 23, no. 2, pp. 724–738, 2015.

[J3] C. Koos, **J. Pfeifle**, and T. Kippenberg, "100 Millionen Telefongespräche über einen Frequenzkamm," *Phys. unserer Zeit*, vol. 45, no. 4, pp. 163–165, 2014.

[J4] **J. Pfeifle**, V. Brasch, M. Lauermann, Y. Yu, D. Wegner, T. Herr, K. Hartinger, P. Schindler, J. Li, D. Hillerkuss, R. Schmogrow, C. Weimann, R. Holzwarth, W. Freude, J. Leuthold, T. J. Kippenberg, and C. Koos, "Coherent terabit communications with microresonator Kerr frequency combs," *Nature Photon.*, vol. 8, no. 5, pp. 375–380, 2014.

[J5] C. Weimann, P. C. Schindler, R. Palmer, S. Wolf, D. Bekele, D. Korn, **J. Pfeifle**, S. Koeber, R. Schmogrow, L. Alloatti, D. Elder, H. Yu, W. Bogaerts, L. R. Dalton, W. Freude, J. Leuthold, and C. Koos, "Silicon-organic hybrid (SOH) frequency comb sources for terabit/s data transmission," *Opt. Express*, vol. 22, no. 3, pp. 3629–3637, 2014.

[J6] J. Leuthold, C. Koos, W. Freude, L. Alloatti, R. Palmer, D. Korn, **J. Pfeifle**, M. Lauermann, R. Dinu, S. Wehrli, M. Jazbinsek, P. Günter, M. Waldow, T. Wahlbrink, J. Bolten, H. Kurz, M. Fournier, J. M. Fedeli, H. Yu, and W. Bogaerts, "Silicon-organic hybrid electro-optical devices," *IEEE J. Sel. Top. Quantum Electron.*, vol. 19, 2013.

[J7] **J. Pfeifle**, L. Alloatti, W. Freude, J. Leuthold, and C. Koos, "Silicon-organic hybrid phase shifter based on a slot waveguide with a liquid-crystal cladding," *Opt. Express*, vol. 20, no. 14, pp. 15359–15376, 2012.

Conference Publications

[C1] J. N. Kemal, **J. Pfeifle**, P. Marin, M. D. Gutierrez Pascual, S. Wolf, F. Smyth, W. Freude, C. Koos, "Parallel multi-wavelength intradyne reception using an optical frequency comb as a local oscillator," in *European Conference on Optical Communication*, 2015, paper P.4.18.

[C2] C. Koos, T. Kippenberg, L. P. Barry, L. Dalton, A. Ramdane, F. Lelarge, J. Leuthold, W. Freude, **J. Pfeifle**, C. Weimann, J. N. Kemal, M. Lauermann, S. Wolf, I. Shkarban, S. Koeber, T. Herr, V. Brasch, R. Watts, D. Elder, "Coherent terabit communications using chip-scale frequency comb sources," *Advanced Photonics Congress, Nonlinear optics,* 2015, paper NTh2A.1.

[C3] W. Freude, **J. Pfeifle**, R. Watts, I. Shkarban, S. Wolf, V. Vujicic, P. Landais, N. Chimot, S. Joshi, K. Merghem, C. Calo, M. Weber, A. Ramdane, F. Lelarge, L. P. Barry, C. Koos, "Phase-noise compensated carriers from an optical frequency comb allowing terabit transmission," 17th Intern. Conf. on Transparent Optical Networks (ICTON'15), Hungary, 2015, paper Tu.B5.1.

[C4] C. Koos, W. Freude, J. Leuthold, M. Kohl, L. R. Dalton, W. Bogaerts, M. Lauermann, A. Melikyan, S. Wolf, C. Weimann, S. Muehlbrandt, **J. Pfeifle**, S. Koeber, R. Palmer, P. C. Schindler, and D. L. Elder, "Silicon-organic (SOH) and plasmonic-organic (POH) hybrid integration: Extending the capabilities of silicon photonics and plasmonics," in *20th OptoElectronics and Communications Conference (OECC'15), Shanghai, China,* 2015.

[C5] **J. Pfeifle**, A. Kordts, P. Marin, M. Karpov, M. Pfeiffer, V. Brasch, R. Rosenberger, J. Kemal, S. Wolf, W. Freude, T. J. Kippenberg, and C. Koos, "Full C and L-band transmission at 20 tbit/s using cavity-soliton Kerr frequency combs," in *CLEO: 2015 Postdeadline Paper Digest*, 2015, paper JTh5C.8.

[C6] **J. Pfeifle**, R. Watts, I. Shkarban, S. Wolf, V. Vujicic, P. Landais, N. Chimot, S. Joshi, K. Merghem, C. Calò, M. Weber, A. Ramdane, F. Lelarge, L. Barry, W. Freude, and C. Koos, "Simultaneous Phase noise reduction of 30 comb lines from a quantum-dash mode-locked laser diode enabling coherent Tbit/s data transmission," in *Optical Fiber Communication Conference*, 2015, paper Tu3I.5.

[C7] **J. Pfeifle**, I. Shkarban, S. Wolf, J. Kemal, C. Weimann, W. Hartmann, N. Chimot, S. Joshi, K. Merghem, A. Martinez, M. Weber, A. Ramdane, F. Lelarge, W. Freude, and C. Koos, "Coherent terabit communications using a quantum-dash mode-locked laser and self-homodyne detection," in *Optical Fiber Communication Conference*, 2015, paper W2A.19.

[C8] C. Koos, W. Freude, J. Leuthold, M. Kohl, L. Dalton, M. Lauermann, A. Melikyan, S. Koeber, S. Wolf[1], C. Weimann, S. Muehlbrandt, K. Koehnle, D. L. Elder, **J. Pfeifle**, R. Palmer, D. Korn, and L. Alloatti, "Silicon-organic hybrid (SOH) and plasmonic-organic hybrid (POH) integration," in *Optical Fiber Communication Conference*, 2015, paper Tu2A.1.

[C9] C. Koos, T. Kippenberg, L. Barry, L. Dalton, W. Freude, J. Leuthold, **J. Pfeifle**, C. Weimann, M. Lauermann, J. Kemal, R. Palmer, S. Koeber, P. Schindler, T. Herr, V. Brasch, R. Watts, and D. Elder, "Terabit/s communications using chip-scale frequency comb sources," *Proc. SPIE* 9343, Laser Resonators, Microresonators, and Beam Control XVII, 93430E, 2015.

[C10] C. Koos, W. Freude, T. J. Kippenberg, J. Leuthold, L. R. Dalton, **J. Pfeifle**, C. Weimann, M. Lauermann, R. Palmer, S. Koeber, S. Wolf, P. Schindler, V. Brasch, and D. Elder, "Terabit/s optical transmission using chip-scale frequency comb sources," in *European Conference on Optical Communication (ECOC)*, 2014, paper Mo.3.4.5.

[C11] C. Koos, T. Kippenberg, L. Dalton, W. Freude, **J. Pfeifle**, T. Herr, C. Weimann, R. Palmer, M. Lauermann, and D. Elder, "Chip-scale frequency comb sources for terabit communications," in *General Assembly and Scientific Symposium (URSI GASS), 2014 XXXIth URSI*, 2014, paper 14693310.

[C12] C. Koos, J. Leuthold, W. Freude, L. R. Dalton, S. Koeber, R. Palmer, C. Weimann, D. L. Elder, W. Heni, D. Korn, **J. Pfeifle**, S. Wolf, D. Bekele, M. Woessner, L. Alloatti, P. C. Schindler, and S. Koenig, "Femtojoule modulation and frequency comb generation in silicon-organic hybrid (SOH) devices," in *Transparent Optical Networks (ICTON), 2014 16th International Conference on*, 2014, paper 14526282.

[C13] V. Vujicic, **J. Pfeifle**, R. Watts, P. C. Schindler, C. Weimann, R. Zhou, W. Freude, C. G. Koos, and L. Barry, "Flexible terabit/s Nyquist-WDM superchannels with net SE > 7bit/s/Hz using a gain-switched comb source," in *CLEO: 2014*, 2014, paper SW1J.3.

[C14] **J. Pfeifle**, V. Brasch, P.C. Schindler, C. Weimann, Y. Yu, T.J. Kippenberg, W. Freude, C. Koos, "Kerr-Frequenzkämme und deren Stabilisierung für die Terabit-Kommunikation," in *115. Jahrestagung der Deutschen Gesellschaft für angewandte Optik*, Karlsruhe, Germany, 2014, paper A2.

[C15] **J. Pfeifle**, Y. Yu, P. Schindler, V. Brasch, T. Herr, C. Weimann, K. Hartinger, R. Holzwarth, W. Freude, T. J. Kippenberg, and C. Koos, "Transmission of a 1.44 Tbit/s data stream using a feedback-stabilized SiN Kerr frequency comb source," in *Optical Fiber Communication Conference*, 2014, paper Th1A.6.

[C16] C. Weimann, P. Schindler, D. Bekele, R. Palmer, D. Korn, **J. Pfeifle**, S. Koeber, R. Schmogrow, L. Alloatti, D. L. Elder, H. Yu, W. Bogaerts, L. R. Dalton, W. Freude, J. Leuthold, and C. Koos, "Data transmission at terabit/s data rates using silicon-organic hybrid (SOH) frequency combs," in *Optical Fiber Communication Conference*, 2014, paper Th4I.2.

[C17] C. G. Koos, W. Freude, J. Leuthold, L. Alloatti, R. Palmer, D. Korn, **J. Pfeifle**, P. C. Schindler, M. Lauermann, N. Lindenmann, S. Koeber, T. Hoose, M. R. Billah, H. Yu, W. Bogaerts, R. Baets, M. Fournier, J.-M. Fedeli, R. Dinu, J. Bolten, T. Wahlbrink, and M. Waldow, "Silicon-organic hybrid integration and photonic wire bonding: Enabling technologies for heterogeneous photonic systems," in *Frontiers in Optics 2013*, 2013, paper FTu1E.3.

[C18] W. Freude, L. Alloatti, D. Korn, M. Lauermann, A. Melikyan, R. Palmer, **J. Pfeifle**, P. C. Schindler, C. Weimann, R. Dinu, J. Bolten, T. Wahlbrink, M. Waldow, S. Walheim, P. Leufke, S. Ulrich, J. Ye, P. Vincze, H. Hahn, H. Yu, W. Bogaerts, V. Brasch, T. Herr, R. Holzwarth, K. Hartinger, C. Stamatiadis, M. O'Keefe, L. Stampoulidis, L. Zimmermann, R. Baets, T. Schimmel, I. Tomkos, K. Petermann, T. J. Kippenberg, C. G. Koos, and J. Leuthold, "Nonlinear nano-photonics," in *Frontiers in Optics 2013*, 2013, paper FTu4C.6.

[C19] C. Koos, W. Freude, J. Leuthold, L. Alloatti, R. Palmer, D. Korn, M. Lauermann, S. Koeber, **J. Pfeifle**, P. C. Schindler, "Silicon-organic hybrid (SOH) integration: Extending the capabilities of silicon photonics," in *Information Photonics 2013 Conference*, University of Warsaw, 2013.

[C20] C. Koos, J. Leuthold, W. Freude, L. Alloatti, R. Palmer, D. Korn, **J. Pfeifle**, P. C. Schindler, and M. Lauermann, "Silicon-organic hybrid (SOH) technology: A platform for efficient electro-optical devices," in *International Conference on Optical MEMS and Nanophotonics*, 2013, pp. 85–86.

[C21] C. Koos, J. Leuthold, W. Freude, L. Alloatti, D. Korn, R. Palmer, M. Lauermann, **J. Pfeifle**, N. Lindenmann, S. Koeber, D. Hillerkuss, R. Schmogrow, "Silicon-organic hybrid integration, photonic wire bonding, and frequency combs: Technologies for multi-Terabit/s interconnects," in *International Nano-Optoelectronics Workshop iNOW-2013*, Cargèse, France, 2013.

[C22] J. Leuthold, C. Koos, W. Freude, L. Alloatti, R. Palmer, D. Korn, **J. Pfeifle**, M. Lauermann, R. Dinu, S. Wehrli, M. Jazbinsek, P. Gunter, M. Waldow, T. Wahlbrink, J. Bolten, M. Fournier, J. M. Fedeli, W. Bogaerts, and H. Yu, "High-speed, low-power optical modulators in silicon," in *Transparent Optical Networks (ICTON), 2013 15th International Conference on*, 2013, paper 13778985.

[C23] **J. Pfeifle**, M. Lauermann, D. Wegner, J. Li, K. Hartinger, V. Brasch, T. Herr, D. Hillerkuss, R. M. Schmogrow, T. Schimmel, R. Holzwarth, T. J. Kippenberg, J. Leuthold, W. Freude, and C. Koos, "Microresonator-based frequency comb generator as optical source for coherent WDM transmission," in *Optical Fiber Communication Conference/National Fiber Optic Engineers Conference 2013*, 2013, paper OW3C.2.

[C24] W. Freude, L. Alloatti, A. Melikyan, R. Palmer, D. Korn, **J. Pfeifle**, C. Weimann, S. Koeber, A. Barklund, R. Dinu, J. Wieland, M. Fournier, J. Fedeli, J. Bolten, T. Wahlbrink, M. Waldow, S. Walheim, P. M. Leufke, S. Ulrich, J. Ye, P. Vincze, H. Hahn, H. Yu, W. Bogaerts, P. Dumont, B. Breiten, F. Diederich, M. T. Beels, I. Biaggio, K. Hartinger, V. Brasch, T. Herr, R. Holzwarth, R. Baets, T. Schimmel, T. Kippenberg, C. Koos, J. Leuthold, "Nonlinear nanophotonic devices," in *Seminar on Optical, Electronic and Quantum Systems, JILA at University of Colorado and National Institute of Standards and Technology (NIST)*, Boulder, USA, 2013.

[C25] C. Koos, J. Leuthold, W. Freude, T. J. Kippenberg, **J. Pfeifle**, C. Weimann, K. Hartinger, V. Brasch, T. Herr, R. Holzwarth, D. Hillerkuss, and R. Schmogrow, "Terabit/s data transmission using optical frequency combs," *Proc. SPIE*, vol. 8600. pp. 860008–860009, 2013.

[C26] L. Alloatti, D. Korn, **J. Pfeifle**, R. Palmer, S. Koeber, M. Baier, R. Schmogrow, S. Diebold, P. Pahl, T. Zwick, H. Yu, W. Bogaerts, R. Baets, M. Fournier, J. Fedeli, R. Dinu, C. Koos, W. Freude, and J. Leuthold, "Silicon-organic hybrid devices," *Proc. SPIE*, vol. 8629. p. 86290P–86290P–12, 2013.

[C27] J. Leuthold, C. Koos, W. Freude, L. Alloatti, R. Palmer, D. Korn, **J. Pfeifle**, and M. Lauermann, "Silicon-organic hybrid - a path towards active silicon photonic devices," in Proc. EOS Annual Meeting (EOSAM 2012), paper 6441, 2012.

[C28] C. Koos, J. Leuthold, W. Freude, L. Alloatti, D. Korn, R. Palmer, M. Lauermann, N. Lindenmann, S. Koeber, **J. Pfeifle**, P. C. Schindler, D. Hillerkuss, and R. Schmogrow, "Silicon-organic hybrid integration and photonic wire bonding: Technologies for Terabit/s interconnects," in Joint Symposium on Opto- and Microelectronic Dvices and Circuits (SODC2012), Hangzhou, China, 2012.

[C29] D. Korn, L. Alloatti, M. Lauermann, **J. Pfeifle**, R. Palmer, P. C. Schindler, W. Freude, C. Koos, J. Leuthold, H. Yu, W. Bogaerts, K. Komorowska, R.

Baets, J. Van Campenhout, P. Verheyen, J. Wouters, M. Moelants, P. Absil, A. Secchi, M. Dispenza, M. Jazbinsek, P. Gunter, S. Wehrli, M. Bossard, P. Zakynthinos, and I. Tomkos, "Silicon-organic hybrid fabrication platform for integrated circuits," in *International Conference on Transparent Optical Networks*, 2012.

[C30] C. Koos, L. Leuthold, W. Freude, L. Alloatti, D. Korn, N. Lindenmann, R. Palmer, **J. Pfeifle**, D. Hillerkuss, and R. Schmogrow, R., "Silicon nanophotonics and photonic wire bonding: Technologies for terabit/s interconnects," in 6th ITG Workshop on Photonic Integration, Laser Optics Berlin, 2012.

[C31] **J. Pfeifle**, C. Weimann, F. Bach, J. Riemensberger, K. Hartinger, D. Hillerkuss, M. Jordan, R. Holtzwarth, T. J. Kippenberg, J. Leuthold, W. Freude, and C. Koos, "Microresonator-based optical frequency combs for high-bitrate WDM data transmission," in *Optical Fiber Communication Conference*, 2012, paper OW1C.4.

[C32] L. Alloatti, **J. Pfeifle**, J. Mendez, W. Freude, J. Leuthold, and C. Koos, "Liquid crystal phase shifter on the SOH Platform with ultra-low power consumption," in *Optical Fiber Communication Conference*, 2012, paper OTu1I.5.

[C33] S. Koenig, **J. Pfeifle**, R. Bonk, T. Vallaitis, C. Meuer, D. Bimberg, C. Koos, W. Freude, and J. Leuthold, "Optical and electrical power dynamic range of semiconductor optical amplifiers in radio-over-fiber networks," in *European Conference on Optical Communication*, 2010, paper Th.10.B.6.

Book Chapter

[B1] K. Yamada, T. Tsuchizawa, H. Fukuda, C. Koos, J. Pfeifle, J. H. Schmid, P. Cheben, and P. J. Bock, "*Handbook of Silicon Photonics*," in: L. Vivien, L. Pavesi (Eds.): Taylor & Francis, Boca Raton, Fl, USA, ISBN 9781439836101, pp. 55-97, 2013.

Curriculum Vitae

Contact

E-Mail: JoergPfeifle@gmx.de

Current Address: Karlstr. 120
76137 Karlsruhe
Germany

Work experience

Feb 2016 – Today	**Keysight Technologies Deutschland GmbH, Böblingen, Germany** Product Manager
Oct 2010 – Dec 2015	**Karlsruhe Institute of Technology (KIT), Karlsruhe, Germany** **PhD candidate** at Institute of Photonics and Quantum Electronics (IPQ) • Characterization and optimization of frequency comb sources • Design and testing of comb-based coherent data transmission experiments with high spectral efficiency • First-time experimental demonstrations of Tbit/s data transmission with three novel frequency comb sources • Establishing and sustaining pan-European collaborations with groups that develop and fabricate the novel comb sources • Participation in several national and international conferences and summer schools; 6 oral and 2 poster presentations • Advisor of several MSc student projects • Teaching assistant: Optical Waveguides and Fibers, Nonlinear Optics, Optical Sources and Detectors, and Semiconductor Devices
May 2010 – Sep 2010	**Research Assistant**, KIT, IPQ Demonstration of an energy-efficient optical phase shifter using liquid crystals on silicon-on-insulator waveguides
May 2008 – Oct 2008	**Internship**, CoreOptics GmbH, Research and Development department Project: Development of a 40G Transponder Tasks: Support setup, control, and automation of testbeds; Design, simulation, buildup, and testing of electrical circuits for the laboratory
Jan 2006 – Jul 2007	Supervisor of the computer tutorial for the lecture 'Digitaltechnik' Tutor for 'Festkörperelektronik' and 'Einführung in die Informatik'

Education

Oct 2010 – Today	**Karlsruhe Institute of Technology (KIT), Karlsruhe, Germany** **PhD candidate** at Institute of Photonics and Quantum Electronics **Supervisor:** Prof. Christian Koos **Topic:** 'Terabit-rate transmission using optical frequency comb sources' PhD program of the Karlsruhe School of Optics and Photonics (KSOP)
14.05.2010	Graduation: **Diplom-Ingenieur**, Grade: 1.5
Sep 2004 – May 2010	**Karlsruhe Institute of Technology (KIT), Karlsruhe, Germany** Study course: **Electrical and Information Engineering** Focus: **Optical Communications** **Topic of master thesis**: 'Semiconductor Optical Amplifier for Radio-over-Fiber Networks'
Sep 2007 – May 2008	**Study abroad** at Queen's University, Kingston, Ontario Ontario/Baden-Württemberg Student Exchange Program Additional Baden-Württemberg Scholarship

Miscellaneous

Feb 2013	KSOP module 'Digital Signal Processing'
Jan 2013	KSOP module 'LabView', including CLAD certification
Nov 2012	KSOP module 'International Project Management'
Aug 2010 – Sep 2010	Long distance hike from Munich to Venice
Oct 2008 – Mar 2010	Participation at AK Erasmus student association Counceling of foreign students studying in Karlsruhe; Director of the sports department
Oct 2008 – Mar 2009	French course: 'Französisch 5', at Sprachenzentrum KIT
Jan 2006 – Jan 2007	English course: 'Professional English for Engineers', VDE
Oct 2004 – Feb 2006	Participation at electrical engineering student association Organization of orientation week for new students and other social events

Additional Qualifications

Languages	German (mother tongue) English (fluent) French (basics)
Software	MATLAB: Regular use for >5yrs LabView: Regular use for >3yrs MS-Office: Regular use for >5yrs Basic knowledge of photonic design and simulation tools: RSoft Photonics CAD suite, VPItransmissionMaker, CST Microwave Studio Basic knowledge of measurement automation tool Keysight VEE